Cambridge Elements ≡

Elements in the Philosophy of Biology
edited by
Grant Ramsey
KU Leuven
Michael Ruse
Florida State University

EVOLUTION, MORALITY AND THE FABRIC OF SOCIETY

R. Paul Thompson
University of Toronto

CAMBRIDGE
UNIVERSITY PRESS

CAMBRIDGE
UNIVERSITY PRESS

University Printing House, Cambridge CB2 8BS, United Kingdom

One Liberty Plaza, 20th Floor, New York, NY 10006, USA

477 Williamstown Road, Port Melbourne, VIC 3207, Australia

314–321, 3rd Floor, Plot 3, Splendor Forum, Jasola District Centre, New Delhi – 110025, India

103 Penang Road, #05–06/07, Visioncrest Commercial, Singapore 238467

Cambridge University Press is part of the University of Cambridge.

It furthers the University's mission by disseminating knowledge in the pursuit of education, learning, and research at the highest international levels of excellence.

www.cambridge.org
Information on this title: www.cambridge.org/9781009244916
DOI: 10.1017/9781108680752

First published 2022

A catalogue record for this publication is available from the British Library.

ISBN 978-1-009-24491-6 Hardback
ISBN 978-1-108-74170-5 Paperback
ISSN 2515-1126 (online)
ISSN 2515-1118 (print)

Evolution, Morality and the Fabric of Society

Elements in the Philosophy of Biology

DOI: 10.1017/9781108680752
First published online: April 2022

R. Paul Thompson
University of Toronto

Author for correspondence: R. Paul Thompson, p.thompson@utoronto.ca

Abstract: Recent interest in the evolution of the social contract is extended by providing a thoroughly naturalistic, evolutionary account of the biological underpinnings of a social contract theory of morality. This social contract theory of morality (contractevolism) provides an evolutionary justification of the primacy of a moral principle of maximisation of the opportunities for evolutionary reproductive success (ERS), where maximising opportunities do not entail an obligation on individuals to choose to maximise their ERS. From that primary principle, the moral principles of inclusion, individual sovereignty (liberty) and equality can be derived. The implications of these principles, within contractevolism, are explored through an examination of patriarchy, individual sovereignty and copulatory choices, as well as overpopulation and extinction. Contractevolism is grounded in evolutionary dynamics that resulted in humans and human societies. The most important behavioural consequences of evolution to contractevolism are reciprocity, cooperation and empathy, and the most important cognitive consequences are reason and behavioural modification.

Keywords: social contract, ethics, morality, sociobiology, naturalism

ISBNs: 9781009244916 (HB), 9781108741705 (PB), 9781108680752 (OC)
ISSNs: 2515-1126 (online), 2515-1118 (print)

Contents

1 Prologetic Remarks

1.1 The Organising Framework

Brian Skyrms opens his excellent 1996 book *Evolution of the Social Contract* thus:

> The best-known tradition approaches the social contract in terms of rational decision. It asks what sort of contract rational decision makers would agree to in a preexisting 'state of nature.' This is the tradition of Thomas Hobbes and – in our own time – of John Harsanyi and John Rawls. There is another tradition – exemplified by David Hume and Jean Jacques Rousseau – which asks different questions. How can the existing implicit social contract have evolved? How may it continue to evolve? This book is intended as a contribution to the second tradition.
>
> (Skyrms, 1996, p. ix)

The evolutionary accounts of Hume and Rousseau were pre-Darwinian and pre-von Neumann and game theory. Skyrms' 'contribution to the second tradition' employs aspects of the theory of games, descended from von Neumann, and evolutionary dynamics, descended from Darwin. This Element brings the full panoply of evolutionary biology to the evolution of the social contract and the justification of the moral theory which emerges therefrom.[1]

Pivotal to this account, rational deliberation occurs late in the evolution of the social contract and not, in the tradition of Hobbes, as generative. The animal propensity for reciprocation, cooperation, intraspecies conflict, territoriality and sociality dates from the Cambrian period, 500 million years ago. For much of that history, there is no capacity for rational decision-making of the kind required by the Hobbesian tradition.

The account of the *justification* of morality developed herein is audacious. Although in many respects it is an extension of a very large body of existing work in ethics, evolutionary biology, evolutionary psychology and anthropology, in important respects it is also a novel approach. Part of its novelty echoes Richard Richards' (2005) statement, 'The concept of fitness, I will argue, is the fundamental valuational concept' (p. 271). His purposes and strategies differ somewhat from mine, but we share common assumptions and perspectives.

---

[1] Whether Hume was a social contract theorist is unsettled. Hume (1772/1984, pp. 186–201), in his essay 'Of the original contract', is critical of the social contract origins of government and state. David Gautier (1979), however, has argued that Hume was a social contractarian, a view that Stephen Buckle and Dario Castiglione (1991, pp. 461–2, n. 15) challenge. What is more settled is that Hume's conception of the origin of society and morals is not based on rational decision-making in a pre-social state of nature; that is what sets him apart from Hobbes, Locke and, more recently, Rawls, Gautier and Harsanyi.

This theory of morality is naturalistic, by which I mean that it is entirely grounded in the natural order; there are no appeals to things or realms beyond what exists in nature and the dynamics underlying the behaviour of what exists in nature. Some versions of naturalism go further and adopt views about the ultimate constituents of nature – monads, for example; the naturalism adopted here is agnostic on these views. Naturalistic stances with respect to moral norms are immediately confronted by some philosophers wielding a version of Hume's 'is/ought' barrier or G. E. Moore's naturalistic fallacy. These objections are dispatched in Section 5.

1.2 Evolutionary Reproductive Success

There is widespread acceptance of the explanatory relevance of biological evolution to morality. Evolution entails that we are animals and, like other animals, have evolved sentiments – empathy, compassion, desire for acceptance, aggression and jealousy, for example – and propensities – reciprocation, cooperation and sociality, for example. We have these because they enhanced survival in some environment in our recent or very distant past. These sentiments and propensities underpin our moral values. Hence, evolution explains the origin of moral values. Also widespread is the view that an explanation of origins does not provide a justification of moral values.

Some ethical theorists claim spiritual (theistic or other credal) access to authentic moral values. These justifications are supernatural – independent of the empirical world. Also, notwithstanding the name, natural law theories of morality are supernatural in that they rely on 'right thinking': that which is obvious to all right reasoning people. 'Natural' here is not synonymous with 'empirical'; something more than (supra) empirical is involved (Haakonssen, 1992; Murphy, 2019).

Non-spiritually based justifications of moral values abound. The history of philosophy overflows with such accounts. In the twentieth century, some theories abandoned the quest for justification altogether. Roger Crisp (2013) provides an excellent collection of articles covering the gamut of theories advanced over the last two and a bit millennia.

The ethical theory developed here is naturalist and contractual and justifies moral principles evolutionarily, hence it is appropriately coined 'contractevolism'. On this theory, biological evolution *explains* how the values we accept have emerged and *justifies* acceptance of them. It draws on a number of fields of empirical inquiry, especially evolutionary biology, psychology, anthropology and sociology.

A key element of the causal mechanisms (the dynamics) of evolution is reproductive success. Elementary reproductive success is the transmission of one's genes into the next generation. Evolution requires more. It requires differential reproductive success (DRS): the transmission of one's genes to the next generation *in greater numbers* than relevant alternate genes. This is the essence of natural selection; 'fitter' organisms will produce more surviving offspring. Of course, like a journey, which is the culmination of many footsteps, so evolutionary change is the culmination of many single generations. Long-term DRS leads to evolutionary reproductive success (ERS), which is the fundamental concept employed here. Superficially, copulatory success is the measure of fitness. As subsequent discussion makes clear, however, copulatory success is not a reliable measure of fitness. A host of other factors are involved: the characteristics of a copulatory partner, available resources, ability and inclination to ensure offspring survival to reproductive viability, the social arrangements that foster or hinder the survival and success of any offspring, for example. In the case of humans, one critical element in nurturing social arrangements is cooperation.

1.3 Reciprocity and Cooperation

Cooperation – mostly rooted in reciprocity – has evolved as a fundamental element of the evolutionary dynamical solution to the maximisation of the ERS of individual organisms. This evolutionary dynamic involves both biological and cultural evolution in higher organisms. Cooperation has resulted in social organisation. A large variety of social structures has arisen through a variety of pathways (see Scarre, 2018; Trigger, 2003). In most cases, in order to survive, a society must function as a coherent whole, be stable and promote the ERS of its members. The warp and weft of a social fabric that supports coherence, stability and ERS include moral principles. I focus on the principles of inclusivity, liberty and equality as essential to reciprocity, cooperation and ERS. There is a continuum, of course; a society, given its circumstances (physical and cultural), might lack maximal inclusivity, liberty or equality, but nonetheless it has more of it than any available alternative social structure under the circumstances of the moment. All things considered, therefore, this is the best alternative at that time, in that place, with the current resources and so on. The underlying assumption here is an evolutionary one; a society with greater inclusivity, liberty and equality will displace one with less because the one with more, on average, will promote more fully the ERS of its members.

1.4 'Evolutionary Ethics'

The account explicated herein has had an almost thirty-year gestation, beginning with an article in the journal *Human Evolution* and later refined in a 1999 article in *Zygon* (Thompson, 1990, 1999). The template for the view explicated here first appeared in *The Monist* (Thompson, 2002). I have benefitted greatly from opportunities to present developing versions to many audiences of evolutionary biologists and philosophers. The comments and penetrating criticisms have been invaluable.

Since 2010, I have avoided the term 'evolutionary ethics'. It has become murky and laden with baggage resulting from more than a century and a half of controversy and considerable confused thinking. The collections of Nitecki and Nitecki (1993) and Thompson (1995) provide a glimpse of this. Decisively, for me, was Robert Richards' assessment, several years ago, that the theory explicated here, albeit in an earlier form, is not evolutionary ethics.

1.5 Deeper Exploration

Given the nature of the Elements series to which this text belongs, I have avoided, for the most part, mathematical formalisms. Those interested in more mathematical treatments of topics such as reciprocity, sexual selection, kin selection and competition will find McElreath and Boyd (2007), Hoppensteadt (1982), Maynard Smith (1974a, 1974b), Bulmer (1994) and Renshaw (1993) valuable. Otto and Day (2007) provide an excellent, comprehensive mathematical approach to biological phenomena. In addition, the length restrictions of the Elements series place constraints on the exposition of the research and analyses of others on which the essence of the theory offered here draws. Hence, in place of such expositions, I have relied on extensive citations of the relevant material. Readers wishing to pursue a topic in more depth can turn to the References.

Furthermore, there are many interesting and important ethical and metaethical issues arising from consideration of evolution and its relevance to morality that are beyond the scope of this Element: moral scepticism, ethical realism and error theory, for example. Those who are interested in exploring these issues cannot do better than to consult Richard Joyce (2007), Richard A. Richards (2005) and John Collier and Michael Stingl (2020).

2 Social Contract Theory

The theory espoused here is a version of social contract theory and it arguably begins with Thomas Hobbes. His was a daring, revolutionary conception of political society and morality. This section provides a potted history to provide a context for evolution-based social contract theory.

2.1 Thomas Hobbes and the Origin of the Social Contract

That *rational* self-interest requires cooperation is one of the fundamental points that Hobbes makes in *Leviathan* (Hobbes 1651).[2] For him, in the state of nature – a state of the world without government and laws – it is every man for himself.

> Hereby it is manifest, that during the time men live without a common Power to keep them all in awe, they are in that condition which is called Warre; and such a warre, as is of every man, against every man. (13:7)
>
> Whatsoever therefore is consequent to a time of Warre, where every man is Enemy to every man; ... consequently no Culture of the Earth; no Navigation, nor use of the commodities that may be imported by Sea; no commodious Building; no Instruments of moving, and removing such things as require much force; no Knowledge of the face of the Earth; no account of Time; no Arts; no Letters; no Society; and which is worst of all, continuall feare, and danger of violent death; And the life of man, solitary, poore, nasty, brutish, and short. (13:8)
>
> To this warre of every man against every man, this also is consequent; that nothing can be Unjust. The notions of Right and Wrong, Justice and Injustice have there no place. Where there is no common Power, there is no Law: where no Law, no Injustice. Force, and Fraud, are in warre the two Cardinall vertues. (13:12)

This is a bleak picture of the condition of man without society, but Hobbes believed that his view conformed to the observational evidence regarding human nature. Human nature, however, is a contested concept, which, historically, has been used to express some perceived *essential*, and usually innate, characteristics of human cognition, motivation or behaviour. I make no use of the concept but find Grant Ramsey's (2013) refreshing and compelling account the most amenable:

> This account of human nature I will label the life-history trait cluster (LTC) account. This is to distinguish it from accounts that are essentialist or normative, since it is neither based on essential properties nor, as we will see in section 7, does it imply that human nature is in any sense 'good.' Instead, characterizations of features of human nature are merely descriptions of patterns within the collective set of human life histories. (p. 988)

This concept of human nature is valuable within an evolutionary context precisely because it is neither essentialist nor normative.

Hobbes' approach to morality is *naturalistic* and it is that feature that remains an important and enduring departure from his predecessors. The fundamental element of Hobbesian human nature is self-interest; this is the cause of the war

---

[2] All references to *Leviathan* are to this edition, but since chapters and paragraphs are referenced (e.g., 13:8), the quoted text can easily be found in any reprinting.

of all against all. The principal self-interest of the person is preservation of life and health against harm. All the other interests of an individual are subservient to this one; physical life and health are the *sine qua non* for all other interests.

A second fundamental feature of Hobbesian human nature is rationality. A rational individual will recognise that sacrificing some secondary interests is the best way to protect her/his life and health. This is the rational motivation for cooperation in the form of an agreement to live in a certain kind of coordinated mutual pact of protection. Rational self-interest is advanced through cooperation; cooperation among a large group of people leads to a society and a social fabric (rules, norms and enforcement). Contemporary evidence, however, suggests that humans have always existed in groups – as do many non-human primates – and cooperation is pervasive among humans (Bissonnette et al., 2015; Voorhees et al., 2020). It is not the result of rationally pursuing self-interest; it is a result of the dynamics of evolution driven by ERS. Those with a propensity to cooperate have greater ERS. This is the cornerstone of the Skyrms tradition and the one developed here.

Hobbes' rational self-interest entails a decision to surrender the pursuit of some lesser interests in order to participate in an organised collective (a society) that guarantees, or at least enhances greatly, the protection of life and health. This, of course, assumes that all other members of the society also have surrendered relevant lesser interests; they have agreed – at least tacitly – to a social contract, the principal feature of which is agreement to live in accordance with the civil laws enacted by the duly acknowledged authority and to accept the designated penalties for failure. One clear obligation that emerges for Hobbes is keeping one's promises: honouring the agreement.

Hobbes entertained three kind of societies that were a rational response to the state of nature: (1) a society governed by a supreme sovereign; (2) one governed by an oligarchy (a small group of rulers); and (3) democracy of some form. Hobbes considers the sovereign model preferable, although he allows that a democracy could be successful. Hobbes' ruler(s) has absolute authority. To disobey the ruler(s), or more aggressively to rebel, is to break one's promise – to dishonour one's agreement. Doing so has the potential to destabilise the society, which puts at risk the protection of life and health of all members of the society, which is contrary to self-interest. This sketches his social contract political theory.[3]

---

[3] Hobbes' embrace of the concept of an absolute authority reflects his historical context. He lived through the English Civil War (1642–51), which was a chaotic period during which the monarch (Charles I) was tried for treason and beheaded. England became a republic (Commonwealth of England). Factions in Parliament then resulted in a protectorate state in which Oliver Cromwell ruled as a de facto dictator, and in 1660 a Convention Parliament declared Charles II the lawful

The dominant moral theory in Hobbes' orbit – natural law – can be traced back to at least Aristotle. Cicero and Thomas Aquinas also advocated versions of it:

> True law is *right reason* in agreement with *nature*: it is of universal application, unchanging and everlasting ... Whoever is disobedient is fleeing from himself and denying his human nature, and by reason of this very fact he will suffer the worst penalties, even if he escapes what is commonly considered punishment. (Cicero, *De Re Publica*, 51 BCE)[4]

> Whatever is contrary to the order of reason is contrary to the nature of human beings as such; and what is reasonable is in accordance with human nature as such. The good of the human being is being in accord with reason, and human evil is being outside the order of reasonableness ... So human virtue, which makes good both the human person and his works, is in accordance with human nature just insofar as it is in accordance with reason; and vice is contrary to human nature just in so far as it is contrary to the order of reasonableness. (Aquinas, *Summa Theologica*, 1265–74)[5]

Reason is inextricably connected to human nature, right action and virtue, such that a failure to live according to the dictates of right reason debases the person (corrupts that person's nature) and, thereby, the goodness of the human person and her/his actions. Aquinas achieved his theologising of natural law by embedding it in God's providence. God has a rational plan for the ordering of all of creation, usually understood as an eternal law, which governs all of creation. Natural law is an element of the eternal law and one of the ways humans, freely and rationally, are involved in the plan of creation. Reason is the human method of discovering it.

Hobbes was critical of this tradition and transformed the concept of natural law while retaining some of its core elements. He rejected Aristotle's conception of natural law and some major elements of Aquinas' conception. He, like Grotius, held that the law of nature was independent of God and discoverable by human reason alone. A key difference between Hobbes and preceding natural law theorists, especially theistically dependent natural law theorists, was that he

---

monarch and England entered a period of parliamentary monarchy – all this in a period of eleven years (1649–60) after the chaos of nine years of war. Hence, Hobbes' passion for stability is understandable. Stability, he argued, requires a ruler(s) that has absolute authority and citizens that have an absolute obligation to obey, an obligation derived from the agreement (contracting) of the individuals within a society to the social structure. Honouring such agreements is rational; it is a law of nature.

[4] The edition cited is Cicero, *De Re Publica*, Cambridge, MA: Harvard University Press, 2006 (Loeb Classical Library, Vol. 16), p. 211, emphasis added.

[5] The edition cited is Thomas Aquinas, *The Summa Theologiæ of St. Thomas Aquinas* (revised ed.), translated by the Fathers of the English Dominican Province, London: Benzinger Brothers, 1920, I–II, q. 71, a. 2c.

held that law and obligation were not a product of reason but could only arise from the command of a sovereign, a view earlier held by Francisco Suárez.

Entangled with Hobbes' conception of natural law is a view of natural rights. The primary natural right was the right to be free, a consequence of which is that a government's legitimacy depends on the consent of the governed. One primary responsibility of those governing is to protect the maximum possible freedom (liberty) of the governed. Hobbes (1651) articulates this primary natural right as:

> the Liberty each man hath, to use his own power, as he will himselfe, for the preservation of his own Nature; that is to say, of his own Life; and consequently, of doing any thing, which in his own Judgement, and Reason, he shall conceive to be the aptest means thereunto. (14:1)

He recognised that laws were a restriction of liberty and as such were in opposition to the natural right of liberty: 'Law is a fetter, Right is freedome, and they differ like contraries' (Hobbes, 1642/1983, 14:3, p. 170).[6] Nonetheless, they are required for social cohesion, which maximises achievable liberty.

2.2 John Locke

John Locke was also an influential social contract theorist. Locke's *Two Treatises of Government* was published a decade after Hobbes' death. He, like Hobbes, uses the state of nature as a starting point. Locke differed from Hobbes on four things. First, Locke believe that a government could be illegitimate and in such cases revolution or insurrection were justified.

Second, Locke held that the equality of each person in the state of nature entails duties one to another, principally justice – not violating the rights of others – and charity. The Golden Rule – 'do unto others as you would have them do unto you'[7] – captures these rights. It is an expression of the principle of reciprocity.

Third, Locke holds that although a condition of no government can exist in the state of nature there also can be government. There can be either a legitimate government – one chosen by a group – but it will, in the state of nature, be limited in power, almost a proto-government, or an illegitimate government, where one person or group dominates over others by force.

---

[6] *De Cive*, which was originally published in Latin, was published in English in 1651 as *Philosophicall Rudiments Concerning Government and Society*, London: R. Royston.

[7] Locke had in mind the Christian version found in Matthew 7:12 (Luke 6:31 has essentially the same command). This echoes 'love your neighbour as yourself' (Leviticus 19:18). Almost all religions (past and present), however, have some version. Many take a command of the Egyptian goddess Ma'at (*c.*2040–1650 BCE) to be a variant, and it is a central tenet of Buddhism.

Fourth, Locke took issue with what he believed to be Hobbes' extreme view of the law as restrictive of liberty. Locke claimed that the primary role of laws was the security of freedom, as stated in a well-known quotation from his *Two Treatises of Government*:

> So that, however it may be mistaken, the end of law is not to abolish or restrain, but to preserve and enlarge freedom: for in all the states of created beings capable of laws, where there is no law, there is no freedom: for liberty is, to be free from restraint and violence from others; which cannot be, where there is no law: but freedom is not, as we are told, a liberty for every man to do what he lists: (for who could be free, when every other man's humour might domineer over him?) but a liberty to dispose, and order as he lists, his person, actions, possessions, and his whole property, within the allowance of those laws under which he is, and therein not to be subject to the arbitrary will of another, but freely follow his own.
>
> (Locke, 1690/1980, p. 32)

Notwithstanding these differences, both accept (1) an initial state of nature (original position); (2) that the primary interest of every individual is survival; (3) that, rationally, government is required to maximise this end of survival; and (4) that the origin and legitimacy of government reside in the consent of those governed.

Locke offers an additional rational basis for government. A violation of the rights of individuals must be prosecuted by the wronged individual. Individuals, however, frequently have an exaggerated perception of the severity of the harm suffered and will mete out a more severe punishment than can be justified by the principle 'punishment should be proportional to harm'. In such cases, an injustice will be done. A government can regulate, prosecute and punish in a disinterested manner, thereby ensuring greater justice.

2.3 Social Contract and Political Revolution

Social contract theory became increasingly important in the eighteenth century, a century of two major revolutions: the American Revolution (beginning in 1775) and the French Revolution (beginning with the storming of the Bastille in 1789). Three influential individuals are noteworthy: Thomas Paine and his interlocutor, Edmund Burke, and Jean-Jacques Rousseau.

Payne and Burke represented two starkly different conceptions of social change. Burke, more in line with Hobbes, held that change most effectively occurs slowly in the context of a stable society. For him, revolution is an abrogation of one's duty to the state and its sovereign. Payne, more in line with Locke, held that revolution was justified when its goal was to replace an illegitimate government. Their views continue to influence political debate today, especially in the United States.[8] A comment in passing, Mary Wollstonecraft in her *Vindication of the Rights of Men* (1790) provided arguably the most effective and insightful critique of Burke's views.

Rousseau's *Social Contract* (*Du contract social*) (1762), the title making clear his theoretical position, was very influential; thirty-two French editions were produced in the decade 1789–99. In the tradition of Hobbes and Locke, Rousseau presented a conception of the state of nature in the *Discourse on the Origin and Foundations of Inequality among Men* (*Un discours sur l'origine et les fondements de l'inégalité parmi les hommes*) (1755). Rousseau, contrary to Hobbes, describes humans in the state of nature as peaceful, alone and lacking a sufficient sense of the future to worry about what is yet to come.

By the end of the eighteenth century, a social contract–based concept of natural rights became codified in the declarations of the revolutionary states. The US declaration (the Declaration of Independence) states:

> We hold these truths to be self-evident, that all men are created equal, that they are endowed by their Creator with certain unalienable Rights, that among these are Life, Liberty and the pursuit of Happiness.[9]

And the French (the Declaration of the Rights of Man and of the Citizen) states:

> **Article first**
> Men are born and remain free and equal in rights. Social distinctions can be founded only on the common good.[10]

2.4 Twentieth-Century Revival

Interest in social contract theory waned during the next two centuries, but in the second half of the twentieth century there was a revival of interest, beginning with John Rawls (1958, 1971). Rawls' theory is principally political but, for him, morality is a consequence of particular political cultures on which a society is based. At the core of a democratic society are three essential factors: first, all citizens are free; second, all citizens are equal; and third, the social fabric exemplifies a fair system of cooperation. These constitute a central feature that places Rawls within the social contract tradition; society is a structure arising from the cooperation of free and equal individuals.

---

[8] See: Yuval Levin's *The Great Debate: Edmund Burke, Thomas Payne, and the Birth of Right and Left*. Although Levin provides a rich account of the views of both men, it is hard to resist concluding that he is biased towards the views of Burke.

[9] See 'Declaration of Independence: A Transcription', National Archives website, www .archives.gov/founding-docs/declaration-transcript.

[10] See 'The Declaration of the Rights of Man and of the Citizen', Élysée website, www.elysee.fr /en/french-presidency/the-declaration-of-the-rights-of-man-and-of-the-citizen.

Rawls, like Hobbes and Locke, has an 'original position', which is a conceptual tool – a thought experiment – in which each individual is free and equal. The goal is to find agreement on principles that are fair. In the original position, no individual has more influence, authority or power than another. Agreement on the principles constitutes a social contract on which the society will be based. Essential to agreement is cooperation and self-interest; rational self-interested cooperators who are free and equal in a *fairly* structured situation – a veil of ignorance about one's circumstances guarantees this in Rawls' thought experiment – will find agreement on principles that become the fabric of a *just* society of free and equal citizens. Fairness is the hallmark of justice.

Another influential version of social contract theory was provided by David Gauthier (1986). His framework is closer to that of Hobbes and his view has become known as contractarianism. Jan Narveson has also advanced a version of social contract theory, which rests principally on the observation that cooperation maximises the security of the person for all participants. His version assumes, as does Hobbes', that, without cooperation and agreements, individuals are exposed to the abuse of others and, hence, risks to their life and bodily integrity. As Narveson (1988) succinctly put it, we agree to a specific social contract 'first because we are vulnerable to the depredations of others, and second because we can all benefit from cooperation with others' (p. 148).

James Buchanan and Gordon Tullock (1962), approaching the issues as economists and in the tradition of social choice theory, have also advanced a sophisticated contractarianism. In place of the state of nature, their 'original position' is the state of 'no collective action'. A fundamental assumption of their version is that individuals will opt for the regulations in a society if the balance of costs favours accepting the regulations over unconstrained freedom, that is, the cost of losing some freedom through regulation is less than the cost of continued unconstrained freedom (with the attendant exposure to predation by fellow humans or other predators). A novel feature of their version is an emphasis on rules for deciding on collective regulation, although, apart from a rejection of decision by unanimity, any specific rule for deciding is problematic. This lacuna, as Michael Lessnoff (1986) points out, is no better resolved in Buchanan's later treatment (Buchanan, 1975). Buchanan was awarded the Nobel Memorial Prize in Economic Sciences in 1986 for 'development of the contractual and constitutional bases for the theory of economic and political decision making'.[11]

---

[11] See 'Press Release', 16 October 1986, The Nobel Prize website, www.nobelprize.org/prizes/economic-sciences/1986/press-release/.

Yet another version of social contract theory was developed by Thomas Scanlon (1998). Scanlon called his version contractualism, a term that had been used loosely by some previous authors but which now is solidly associated with Scanlon's version and the versions of others sharing its essential assumption, that is, legitimate principles in a society are those that no member can *reasonably* reject. As this indicates, Scanlon focusses on 'reasonable rejection' where most other social contract theorists focus on 'rational acceptance'. Scanlon believes using non-rejectability as the criterion of a legitimate principle provides a direct reason for people to put themselves in other people's positions; reflecting on what other people might reasonably reject will produce greater insight into what other people will accept as a principle than will reflecting *directly* on what others might accept. Scanlon's contractualism rejects aggregation such as 'greatest good for the greatest number'; Parfit's (2011) counter-examples to this rejection are compelling.

In general, contractualists hold that respecting persons is a rational requirement. Respect-for-persons logically entails that justifiable principles in society are those that each person accepts as reasonable. It is not self-interest that underpins agreement on principles but the need for each individual to be convinced that the principles are justified – namely, convinced that respecting other individuals entails, as a strategy of justification, placing oneself in the shoes of others (empathy). This is well captured in the ancient maxim 'do unto others as you would have them do unto you', which was cited in Section 2.2 and plays an important role in an evolutionary-based theory of morality.

A concluding observation: For Rawls, there is a connection between his social contract and evolutionary theory. His social contract theory rests on psychological laws, which are the legacy of evolutionary dynamics. As Rawls (1971) expresses it:

> In arguing for the greater stability of the principles of justice I have assumed that certain psychological laws are true, or approximately so . . . [O]ne might ask how it is that human beings have acquired a nature described by these psychological principles. The theory of evolution would suggest that it is the outcome of natural selection; the capacity for a sense of justice and the moral feelings is an adaptation of mankind to its place in nature. As ethologists maintain, the behavior patterns of a species, and the psychological mechanisms of their acquisition, are just as much its characteristics as are the distinctive features of its bodily structures; and these patterns of behavior have an evolution exactly as organs and bones do. It seems clear that for members of a species which lives in stable social groups, the ability to comply with fair cooperative arrangements and to develop the sentiments necessary to support them is highly advantageous, especially when individuals have a long life and are dependent on one another. These conditions guarantee

innumerable occasions when mutual justice consistently adhered to is beneficial to all parties. (pp. 502–3)

Rawls looks to biological evolution to explain and underpin certain psychological laws assumed by his account of justice. For him, evolution does not
'justify' moral claims; it provides an account of the context within which the
justification of moral norms can be framed.

3 The Evolution of Social Behaviour and Sociality

3.1 Evolutionary Dynamics

Modern evolutionary theory descends from Charles Darwin's 1859 landmark
volume *On the Origin of Species*. In it, Darwin appealed to the powerful
metaphor of 'natural selection' and, influenced by Herbert Spencer, the idea
of 'survival of the fittest'. Although the metaphor of natural selection is pervasive in expositions of evolutionary theory, contemporary evolutionary theory is
cast in terms of 'differential reproductive success' (DRS). For an organism to be
reproductively successful, its physiology and/or behaviours must *influence* the
transmission of its genes to the next generation, specifically the transmission of
those genes that underlie the physiology or behaviours that led to its reproductive success. Living for 100 years is a different kind of survival and success. For
an organism to be *differentially* reproductively successful, it must *outperform*
organisms with alternate physiologies and behaviours in the transmission of its
genes into the next generation. To be evolutionarily relevant, DRS must continue for many generations, a reproductive success hereafter called evolutionary
reproductive success (ERS), which is understood to be long-term DRS. Clearly
ERS depends on the survival of the offspring which carry the parental genes.
Hence, ERS is dependent on physiological and behavioural characteristics that
enhance success in reproduction *and* in offspring survival.

In order for an organism to transmit her genes to the next generation, those
genes must be heritable. Population genetics, quantitative genetics, transmission genetics and molecular genetics provide a clear understanding of the
mechanism of heredity (Graur and Li, 2000; Hartl and Clark 2007; Lynch and
Walsh 1998). Also, for there to be *differential* reproductive success, there has to
be relevant differences among organisms; that is, there must be variation in the
population. In the absence of mechanisms to generate or maintain variation,
a population will experience a constant diminution of that variation because
DRS will cause an increase in each generation of those organisms with the more
successful genomes. Hence, a population will move to a state called fixation,
where all organisms are genetically identical in those respects that made them
reproductively successful, at which point evolution would cease. Evolution,

however, has continued unabated for hundreds of millions of years. Hence, there must be mechanisms that constantly create new varieties. It is now known that there is a lot of latent genetic variability in a population, which through the recombination of genes, during sexual reproduction, yields novel characteristics. Also, mutation produces novel characteristics. Mutations are changes in the genetic code caused by, among other things, environmental mutagens such as radiation and errors in replication. Only mutations that affect the gametic genes (egg and sperm) are transmissible; somatic mutations – many cancers, for example – are not heritable even though in some cases a susceptibility to such mutations might be gametic and heritable. Some mutations are silent; they have no relevant effect on the phenotype (physical and behavioural properties of the adult). Some are manifest in the phenotype but are neutral with respect to ERS. Most are deleterious, negatively affecting the phenotype's viability. A few are advantageous to the organism's viability and ERS. Mutations occur at a very high rate, so even a very few viable ones are more than enough to keep levels of variability high. Frequency-dependent selection is also a mechanism that maintains variability (Hartl and Clark, 2007; Wright, 1969: chap. 5). There are two kinds of frequency-dependent selection: positive (as the frequency of a variant increase, its fitness increases) and negative (as the frequency of a variant increases, its fitness decreases). The latter seems more common and there is a wealth of examples (Frank, 2000). In essence, at a certain frequency a variant within a population is in equilibrium. Below that frequency, selection favours its increase; above that frequency, its selective advantage is lost. Also, as Sarah Otto and colleagues (2008) have shown, 'Frequency-dependent selection is commonly considered in speciation models because it can, under the right circumstances, generate disruptive selection while maintaining a polymorphism' (p. 2092). A variation introduced by John Maynard Smith (1982) uses game theory to explain the equilibria states of interacting organisms (see also Kokko et al., 2006). Frequency-dependent selection appears again in Section 4.7 in this Element, where the problem of 'free-riders' is discussed.

An organism's ERS is always relative to a certain environment. That environment includes other organisms of the species (competitors for resources, for example), other species (predators, for example), physical conditions (climate and terrain, for example) and the like. Change the environment, sometimes very little, and formerly advantageous characteristics may be disadvantageous or neutral to ERS – the first may become last and the last become first.

ERS, heredity, variability and environment are components of the dynamics of evolutionary change. As in physics where the behaviour of a ball on an inclined plane yields important knowledge, different areas of evolutionary research focus on *components*, not the entire system. Of course, a massive landslide is complex as is evolution in a multi-organism environmental context. Although the various factors in a landslide can be explored individually, to understand the complexity of a landslide requires the application of a concatenation of factors (forces, objects' sizes, density, location and fragility, for example). This concatenation of factors constitutes the dynamics (how things behave) and ontology (what entities there are) of a complex system. A theory is a composite description of the entire dynamics and ontology. The General Theory of Relativity is a composite description of this kind as is evolutionary theory; indeed, both are well-confirmed descriptions with robust explanatory and predictive power.

From an evolutionary point of view, it is genes that are transmitted from one generation to another, and DRS determines which genes are transmitted the most successfully. How genes are transmitted, however, is far from simple. The various fields of genetics explore and codify how genes are replicated, transmitted, recombine and mutate. These dynamics for higher organisms like humans are complicated. Quantitative traits (or multifactorial traits) are those that are determined by complexes of genes – not single genes – and have a high degree of environmental plasticity; they are Mendelian but do not obey simple Mendelian transmission (Lynch and Walsh, 1998). A person's height, for example, can be significantly affected by nutrition – an environmental factor. A change in any component of the complex can result in a significant change in the phenotypic trait. Some quantitative traits are 'meristic': they are discrete and countable (number of fingers, for example). Some are 'continuous': they vary across the members of a population over a continuous range of values (height, weight, hand size, brain size, for example). These typically conform to a Gaussian distribution, also known as a normal curve/distribution or, because of its shape, a bell curve.

Epigenetic factors further complicate the transmission process. Epigenetics explores and codifies factors that can change how genes are expressed and result in different phenotypes without changes to genes themselves – factors such as environmental impacts, the effects of some genes (regulatory genes) on the expression of others and temporal embryological (developmental) sequences.

Some organisms – many plants and microbes, for example – sometimes or always reproduce asexually (vegetative propagation, budding, fragmentation and spore production, for example). This is a conceptually rich area of evolutionary genetic research. Humans, however, reproduce sexually; each mating male and female contributes one-half of the genes of the new organism. Humans are the focus of our interests. Hence, sexual reproduction is the relevant evolutionary domain.

3.2 Sociobiology

Animal behaviour has been studied for centuries but the origin of a scientific approach dates from the seventeenth century with the works of Charles-Georges Le Roy and others (Egerton, 2016). The study of animal behaviour today (also called ethology) embraces the genetic, developmental, physiological and neurological bases of behaviour. The study of *social* behaviour focusses on the behaviour of animals in groups. Sociality, as used here, is the propensity of individuals in a population to behave as an organised collective: to work together to achieve a common goal. The organised behaviour of 'social' insects (ants, termites, bees and the like) is a remarkable example. These 'eusocial' organisms have strict roles within the group and the survival of the group and, hence, each organism depends on the others. Non-human primate 'social' groups are less rigidly organised and more multifaceted.

The evolutionary study of human social behaviour, which is pivotal to an evolution-based social contract moral theory (contractevolism), is fraught with dangers. Understandably, humans, since before Darwin and definitely after him, have been touchy about their place in the animal kingdom; human exceptionalism has been bruised many times by science. Sigmund Freud eloquently captured the impact of the advancement of science on a human-centric conception of the universe in his 1917 *Introductory Lectures on Psychoanalysis*:

> In the course of centuries the *naïve* self-love of men has had to submit to two major blows at the hands of science. The first was when they learned that the earth was not the centre of the universe but only a tiny fragment of the cosmic system of scarcely imaginable vastness. This is associated with Copernicus ... The second blow fell when biological science destroyed man's supposedly privileged place in creation and proved his descent from the animal kingdom and his ineradicable animal nature ... But human megalomania will have suffered its third and most wounding blow from the psychological research of the present time which seeks to prove to the ego that it is not even master in its own house, but must content itself with scanty information of what is going on unconsciously in its mind. (Freud, 1971)

In 1975, the entomologist Edward O. Wilson added to the injury. In *Sociobiology: The New Synthesis*, he explored, and promoted, explanations of the behaviour of organisms (including humans) that are grounded in evolutionary theory (Wilson, 1975). His goal was to demonstrate that all aspects of animal behaviour have been influenced by the evolutionary lineage of a species. It provoked a storm of controversy. Leading critics were Marshall Sahlins, Richard Lewontin, Stephen Jay Gould, Richard Levins and a group called Science for the People. It was regarded as genetically deterministic, too adaptationist and dismissive of learning and culture. The criticisms undeniably identified weaknesses, especially when Wilson's focus was on cognitively capable animals. The message of nuanced critics is sound; in the case of humans, there is no place for genetic determinism, claims of adaptation require scrutiny and, importantly, learning and culture are essential elements in human evolution and behaviour. Although true of many other animals as well (Kitcher, 1985), humans are our principal concern. Even acknowledging all of this, however, the vociferousness of the criticisms was disproportionate; the signal-to-noise ratio was often low.

Consider Marshall Sahlins' (1976) criticisms. He claimed that non-kin adoption in Oceania (Polynesia in particular) was high and could not be explained by evolutionary biology (i.e., sociobiology). Shortly after Sahlins' book appeared, Joan Silk (1980) analysed extensive data on large samples of adopters and adoptees in eleven different Oceania cultures, concluding:

> Here I will argue, contra Sahlins, that genetic relatedness is a fundamental, albeit not necessarily conscious, consideration in adoptive decisions, and that details of adoptive behavior are uniformly consistent with sociobiological predictions. The argument is developed from a qualitative model which illustrates the effects of genetic relatedness and family size upon adoptive decisions, and provides the basis for a number of specific predictions regarding the form of adoptive behavior. These predictions are then evaluated with detailed data culled from the extensive ethnographic literature on adoption in Oceania. The results of this analysis indicate quite clearly that adoption facilitates adjustment of family size, and that relatedness is related to the selection and treatment of adopted children and the jural nature of adoptive relationships.
>
> It should be noted from the outset that the aim of this paper is to demonstrate that Oceanic adoption practices are consistent with predictions derived from sociobiological theory. Documentation directed toward this point does not imply that adoptive behavior can be fully explained without reference to culture and ecological conditions. Consideration of these factors, however, lies beyond the scope and intention of this paper. (p. 801)

She concludes that Sahlins' analysis was incomplete and his conclusion unsupported by the data. Why would a seasoned, respected anthropologist be so careless? Attributing motives is a mug's game, so caution is in order, but ulterior motives, conscious or unconscious, cannot be dismissed cavalierly. One such motive, I believe, is a legitimate concern that claims about the evolution of human behaviour have social and political implications; they affect the lives of individuals; they feed ideologies. This is a sobering observation, one that elevates the duty of care in scientific enquiry and conclusions, but not one that entails that the study of the evolution of human social behaviour is off limits.

What is in a name? The intense controversy over *Sociobiology*, regrettably, led many to recast and rename their work as behavioural ecology. Sociobiology, however, is more comprehensive than ecology. Moreover, during the two decades following *Sociobiology*, exploration of the neurocognitive determinants of evolved behaviours gave rise to evolutionary psychology. Frans de Waal (1997) claimed his work to be in another – at that time incipient – field, cognitive ethology, which Marc Bekoff (1995) has defined as, 'the comparative, evolutionary, and ecological study of nonhuman animal (hereafter animal) minds including thought processes, beliefs, rationality, information-processing, and consciousness'. Alan Kingstone (2008) has extended it to humans. These all fall under the broader umbrella of sociobiology. That, like many emerging areas of research, it got off to a rocky start hardly seems a rational justification for jettisoning the term or the scope of the endeavour. One finds, for example, in the works of Sarah Blaffer Hrdy (1981, 2000, 2011) an excellent display of the breadth and depth of sociobiology as it has matured.

3.3 Game Theory and the Evolution of Cooperation

'He had two convictions which persisted throughout his scientific career . . . The first was that, if you are faced by a difficulty or a controversy in science, an ounce of algebra is worth a ton of verbal argument', claimed Maynard Smith (1965) approvingly about J. B. S. Haldane. I agree and have argued that mathematics is the language of science (Thompson, 2007, 2011, 2014).

One branch of mathematics employed in evolutionary biology is game theory, developed by John von Neumann and Oskar Morgenstern (1944) and applied initially to economics. George Price was the first to see a game theoretical solution to behavioural interactions and to have identified stability as a key component of the evolution of many behaviours. In an unpublished paper (Price, 1969), he explored non-lethal conflict within a species.[12] Later, in the hands of Maynard Smith and Price (1973), this behavioural stability in biological species was named 'evolutionary stable strategies' (ESS). Their paper addressed intra-species non-lethal conflict in the context of the debate about group selection versus individual selection (see Sober, 1994; Sober and Wilson, 1998). Although Maynard Smith was open to group selection in some very limited and special cases, in this paper they argued that invoking group selection is unnecessary in explaining intra-species non-lethal conflict; their explanation is game-theoretic.

---

[12] The paper was accepted by *Nature* on 7 February 1969 on the condition that it was to be shortened, which never occurred.

Price extended his investigation to the evolution of cooperation:

> The literature he [Price] was reading was now beginning to finally settle in his head. George C. Williams, he now discovered, had made a suggestion on the matter in his 1966 book *Adaptation and Natural Selection*: Animals that live in stable social groups and that are intelligent enough to form personal friendships and animosities beyond the limits of family could evolve a system of cooperative behavior. But the evolution of reciprocity, Price now saw, actually demanded much less: In a species where cooperative behavior is important, the logic of games would suffice to ensure cooperation. There was really no need for the ability to form friendships and hates; the trick, rather, was for noncooperative behavior to be retaliated against. (Harman, 2011, p. 3)

Robert Trivers (1971) employed game theory to mathematically underpin his account of an aspect of animal cooperation – reciprocal altruism. The most prominent employers of game theory to social contract theory and evolution are the philosopher Brian Skyrms, the biologist John Maynard Smith and the economist Kenneth Binmore.

In the last two decades, the limitations of game-theoretic explanations of behaviour have become more apparent. As Sarah Otto and Troy Day (2007) have noted:

> Game-theoretic models are based on the very same principles of evolutionary invasion analysis presented in this chapter, although many such models do not have an explicit genetic model underlying them or even an explicit population-dynamic model. The vast majority of game-theoretic models have been developed to understand the evolution of social interactions and have consequently ignored many of the genetic and ecological details of the species in question. Instead these models focus on other quantities besides the number of individuals or genes (e.g., the amount of some desirable quantity, such as money or spare time). As this chapter has demonstrated, however, we can consider the spread of alternative strategies across a broad array of genetic and ecological contexts. (p. 498)

John McNamara (2013) makes a similar point:

> The ESS condition is concerned with determining whether a population adopting a particular resident strategy π^* can be invaded but says nothing about whether strategy π^* will evolve in the first place. To analyze this latter question we first have to specify how the resident strategy changes as the population evolves; in other words, we have to specify the evolutionary dynamics. The exact dynamics depend on assumptions about the underlying genetic system; however, one especially simple approach known as adaptive dynamics ignores genetic detail and assumes that selection acts to locally hill climb. (p. 628)

Both genetics and selective pressures at the phenotypic level as studied using game theory are important components of evolutionary dynamics; game theory gives us part of the picture. Understanding the evolution of a characteristic always requires the genetic component.

Modern evolutionary accounts of animal social behaviour are often dated from W. D. Hamilton's (1964a, 1964b) articles on the genetical evolution of social behaviour. In those articles, he extended Sewall Wright's (1922) coefficient of inbreeding and relatedness to explain the fitness implications of relatedness. The closer the relationship of two individuals, the more genes they have in common, where 'in common' means 'have identical copies of a gene'. Identical twins have 100 per cent of their genes in common. Siblings with the same father and mother have 50 per cent of their genes in common. The degree of genetic relatedness is equal to the probability that two individuals will have a relevant gene in common. Since children inherit 50 per cent of their genes from the mother and 50 per cent from the father, the probability that two children will have a particular gene in common is 50 per cent – or more technically 0.5. The importance of this, as Hamilton demonstrated, is that if one of two sisters dies before reproducing but the other has offspring, the DRS of the non-reproducing sister is 50 per cent of that of her reproducing sister. Maynard Smith (1964) coined the term 'kin selection' for this effect of genetic relatedness.

That relatedness is evolutionarily significant was appreciated before Hamilton. John Maynard Smith (1975) wrote that J. B. S. Haldane, 'who had been calculating on the back of an envelope for some minutes, announced that he was prepared to lay down his life for eight cousins or two brothers'. Darwin also commented on relatedness in a way that once again demonstrated his care and brilliance, coming extremely close to the concept of kin selection in *Origin*:

> But with the working ant we have an insect differing greatly from its parents, yet absolutely sterile; so that it could never have transmitted successively acquired modifications of structure or instinct to its progeny. It may well be asked how is it possible to reconcile this case with the theory of natural selection?
>
> . . .
>
> This difficulty, though appearing insuperable, is lessened, or, as I believe, disappears, when it is remembered that selection may be applied to the family, as well as to the individual, and may thus gain the desired end. Thus, a well-flavoured vegetable is cooked, and the individual is destroyed; but the horticulturist sows seeds of the same stock, and confidently expects to get nearly the same variety; breeders of cattle wish the flesh and fat to be well marbled together; the animal has been slaughtered, but the breeder goes with confidence to the same family. (Darwin, 1859, pp. 237–8)

Hamilton generalised this insight and gave it mathematical expression, which allowed it to be integrated into population genetics. Since its conception, kin selection

> has shed light on a range of biological phenomena, including dispersal, sex-ratio adjustment, worker–queen conflicts in insect colonies, the distribution of reproduction in animal societies (reproductive skew), parasite virulence, genomic imprinting, and the evolution of multicellularity (Bourke 2011[...]). The principles of kin selection also help illuminate aspects of the major transitions in evolution, which occur when free-living individuals coalesce to form a new higher-level entity that eventually becomes an individual itself (Maynard Smith and Szathmáry 1995, Bourke 2011[...]). (Birch and Okasha, 2015, p. 22)

The importance of kin selection for contractevolism resides in its partial explanation of the evolution of cooperation. The emergence of a gene (or gene complex) predisposing an organism to cooperate will confer on cooperating relatives a selective advantage – the more closely related the greater the advantage. Each member of a family, for example, will have a DRS based on her number of offspring (primary fitness) but, through cooperation, she can add some portion of the DRS of relatives to her own (secondary fitness via kin selection). The total fitness of an individual is her 'inclusive fitness'. A gene for cooperation enhances the kin selection portion by increasing the probability that a relative will have more offspring with her cooperation than without. Hence, a gene for cooperation can confer a selective advantage on the possessor, making it likely that the gene will increase over the generations. When the gain to an individual's secondary fitness from cooperation is greater than the cost to her primary fitness, cooperation is selectively advantageous and the number of individuals with the genetic propensity will increase in subsequent generations.

Superficially, an individual who increases the reproductive success of her sister at the expense of her own appears to be 'altruistic'. Hamilton, and others since, have claimed that kin selection explains altruism. The concept of altruism, however, has a long history and, as Elliott Sober (1994) has pointed out, its use in this biological context is conceptually distinct from other more conventional uses:

> The concept of altruism has led a double life. In ordinary discourse, as well as in psychology and the social sciences, altruism refers to behaviors that are produced because people have certain sorts of motives. In evolutionary biology, on the other hand, the concept is applied to behaviors that enhance the fitness of others at expense to self. (p. 8)

This thorny linguistic issue, however, need not concern us; all that is needed is an explanation of the evolution of cooperation, of which kin selection provides a partial explanation – partial because cooperation among close relatives does not account for the more general propensity to cooperate, and it is this more general propensity that is needed to undergird a social contract among a large group of cooperating individuals with distant familial relationships at best.

In a seminal paper, Robert Trivers (1971) employed game theory to explain this more general kind of cooperation. The central concept is reciprocity – essentially, you help me and I will help you. Trivers drew on a specific game known as the prisoner's dilemma. As with all behaviours, there are costs and there are benefits.

Imagine two friends commit a major crime. They both agree that, if caught, they will remain silent. They are caught but the police only have enough circumstantial evidence to convict them on a minor offence, which has a jail term of one year. They are interrogated in different rooms. If they remain silent, they both get a one-year sentence. The police offer a deal to each; if you talk and your partner doesn't, you go free and your partner gets ten years. If, however, you both talk, we offer you a reduced sentence of five years (see Figure 1).

If this is a one-shot deal, cooperation means that you need to have remarkable confidence in your partner, because not talking might cost you ten years in prison. The weaker your confidence, the greater the impulse to cut your losses and defect. If your partner remains quiet, you go free; if not, you get five years. If, however, you remain silent and she doesn't, you get ten years and she goes free. Seems rational to talk. But your partner will do the same calculation. Hence, there is a high probability that you'll both get five years.

If, however, you get to play many games of 'cooperate or defect' with this partner, perhaps with varying costs and benefits, the person who chooses his

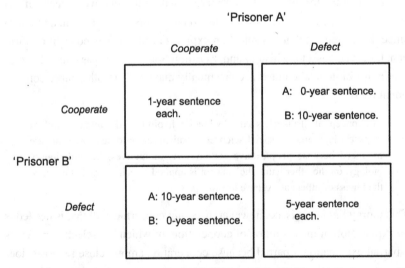

Figure 1 The prisoner's dilemma: the payoff matrix

next decision based on how her partner behaved the time before (tit-for-tat) does better than someone who always defects (Axelrod and Hamilton, 1981). More recently, Nowak and Sigmund (1993) argued that 'win-stay, lose-shift' outperforms 'tit-for-tat'. The theoretical conclusion drawn from the application of this game-theoretic framework to the evolution of behaviour is that reciprocators do better than non-reciprocators over a lifetime of encounters (iterative games). Stephens (1996) states the issue succinctly:

> Roughly, reciprocal altruism evolves because organisms do better by accepting the immediate costs of helping another organism in order to reap the comparatively greater benefits of receiving help at a later time.
> In order to develop this idea precisely, biologists rely extensively on the two-person iterated Prisoner's Dilemma game (Trivers [1971]; Axelrod and Hamilton [1981]; Maynard Smith [1982]; Axelrod [1984]; Dugatkin [1988]; Nowak and Sigmund [1993, 1994]). (p. 534)

Considerable research (theoretical and empirical) followed Trivers' article; Robert Axelrod (1984) summarises the culmination of the early research.

Stephens, however, also argues – convincingly, in my view – that the standard iterated prisoner's dilemma is neither necessary for modelling reciprocity nor without aberrant outcomes:

> In this paper I argue that the iterated Prisoner's Dilemma game is but one way formally to model claims about reciprocal altruism. The standard iterated Prisoner's Dilemma game analysis makes an unnecessary assumption, and furthermore the Prisoner's Dilemma game is itself unnecessary to formally represent reciprocal altruism. I will proceed roughly as follows: first, I develop carefully the intuitive, *informal* conditions necessary for reciprocal altruism; second, I argue that both Axelrod's ordering condition and the anti-exploitation condition are unnecessary. Next I show that other kinds of models, which I call the *modified* Prisoner's Dilemma and *Cook's Dilemma* games meet the informal conditions on reciprocal altruism and therefore can be used to model certain kinds of reciprocal altruism. Then I apply these alternate games to examples and briefly explore how the stability dynamics in the alternate models differ from the standard model. Finally, I conclude with reference to some general issues in adaptationism. (p. 536)

His *cook's dilemma* (as in too many cooks spoil the broth) captures better delayed cooperation and avoids cases of disastrous *simultaneous* cooperation, which are not avoided by the standard prisoner's dilemma game (p. 540).

Another strand of research adds population structure to iteration and delayed benefits (Bergstrom, 2003; Eshel and Cavalli-Sforza, 1982; Grafen, 1985; Nowak et al., 2010; Tarnita et al., 2009; van Veelen et al., 2012). Van Veelen and colleagues (2012) modelled a population structure called association.

Individuals do not meet at random; players are matched with an individual that uses cooperation a percentage of the time, where that percentage is the cooperation frequency in the population. Mathematically: $\Pr(\alpha + (1-\alpha)x_s)$, where x_s is the frequency of cooperation in the population and α is the assortment parameter (the probability that a rare mutant meets this strategy). They conclude:

> [W]e have shown that repetition alone is not enough to support high levels of cooperation, but that repetition together with a small amount of population structure can lead to the evolution of cooperation. In particular, in the parameter region where repetition is common and assortment is small but nonzero, we find a high prevalence of conditionally cooperative strategies. These findings are noteworthy because human interactions are typically repeated and occur in the context of population structure. (p. 9933)

That is the state of play in the theoretical research. With respect to the empirical research, it is extensive. A meaningful review is beyond the scope of this Element, but a significant glimpse can be gained from overviews such as Dugatkin (1997), Davies, Krebs and West (2012), Kasper (2017) and Cheney (2011), and the references given therein. Sufficient for the purposes here is the clear validation of the view that, on theoretical and empirical grounds, the evolution of reciprocity, cooperation and sociality, especially in humans, is well established and theoretically explainable.

3.4 A Caveat

It is important to recognise that, although evolutionary theory explains the human propensity to reciprocate, cooperate and socialise, that propensity is not equally distributed, as van Veelen and colleagues' (2012) model mentioned in the previous section takes seriously. Some animals in a species will be very weak reciprocators and cooperators, some will be exceptionally robust, but most will be average. Mathematically, the distribution will be Gaussian (normal or bell-shaped) in conformity to the central limit theorem: whenever variation in a characteristic is the result of the cumulative effect of a large number of factors, the variation will conform to a normal distribution.

It is also important to recognise that there is much left to learn about the evolution of sociality and its fitness consequences. As Joan Silk (2007) states:

> This general argument implies that sociality has fitness consequences for individuals. However, for most mammalian species, especially long-lived animals like primates, there are sizable gaps in the chain of evidence that links sociality and social bonds to fitness outcomes. These gaps reflect the difficulty of quantifying the cumulative effects of behavioural interactions on

> fitness. This problem is common to almost all studies of the adaptive function of social behaviour in animals. Instead, we generally rely on what Alan Grafen (1991) called the 'phenotypic gambit', the assumption that the short-term benefits that individuals derive from social interactions are ultimately translated into long-term differences in fitness. For example, if group size reduces vigilance time, then individuals will be able to forage more efficiently, and enhanced foraging efficiency will be ultimately transformed into fitness gains. Similarly, we assume that animals which are regularly supported in antagonistic confrontations or groomed frequently gain short-term benefits that enhance their lifetime fitness. Social relationships that provide these kinds of short-term benefits are therefore assumed to have selective value for individuals. This logic is sometimes extended one step further. It is hypothesized that the magnitude of the investment that animals make in their social relationships provides a measure of their adaptive value (Kummer 1978). This hypothesis cannot be tested without information about the adaptive consequences of social bonds. (p. 540)

More work is indeed needed, an enduring truth in all sciences. Nonetheless, the *fact* of the propensity is clear and the broad outlines of the dynamics are well accepted. It is more research on the fitness advantages for which Silk identifies a need, not more evidence of the phenomenon itself. Moreover, as the 'gaps in the chain of evidence' have been filled, they so far have justified the well-accepted dynamical accounts and the significant fitness gains from sociality.

4 Contractevolism: An Evolution-Based Moral Theory

4.1 An Early Attempt

Contractevolism embeds morality within the fabric of a society: a social contract. The underlying principles are biologically based. They are justified, principally, by evolutionary dynamics not, as with contrarianism, by rational agreement.

Let's begin with a fairly unadorned and naïve, but succinct, statement of the evolutionary approach to justifying moral claims. Edward O. Wilson (1975) opens his book *Sociobiology: The New Synthesis* with:

> Camus said that the only serious philosophical question is suicide. That is wrong even in the strict sense intended. The biologist, who is concerned with questions of physiology and evolutionary history, realizes that self-knowledge is constrained and shaped by the emotional control centres in the hypothalamus and limbic system of the brain. These centers flood our consciousness with all the emotions – hate, love, guilt, fear, and others – that are consulted by ethical philosophers who wish to intuit the standards of good and evil. What, we are then compelled to ask, made the hypothalamus and limbic system? They evolved by natural selection. (p. 1)

The essential components of his argument are as follows:

1. The hypothalamus and the limbic system evolved by means of differential reproductive success (natural selection) during a very long period of time and through the lineages of a large number of non-human organisms.
2. The hypothalamus and the limbic system are the locus of all human emotions (disgust, hate, love, empathy and the like).
3. Human emotions are, therefore, no more than the response of an organism to its environment – a response that is a function of its natural biological origins.
4. Moral intuitions about good and evil, right and wrong, are merely reflections of these emotions; they are grounded in nothing more than this.

Wilson points us in the right direction, but his account is underdeveloped and far too narrow. For one thing, contractevolism requires more than biology. Philosophy, sociology, anthropology, political economy and history have much to contribute. For another, although Wilson subsequently provided more nuanced accounts (Lumsden and Wilson, 1981; Ruse and Wilson, 1986; Wilson, 1978), his account here is unnecessarily biologically deterministic; it underplays the role of cognition. Many human propensities, such as the emotions he mentioned, are powerful and deeply embedded in our evolved natures but are also frequently tempered by cognition. Wilson focussed on the limbic system, the structures of which are located within the brain, underneath the cerebral cortex and above the brainstem. It generates our emotional and behavioural responses – those we need for survival: feeding, reproduction and caring for offspring. Two components of the limbic system (the thalamus and hypothalamus) produce hormones important for survival (hunger, mood, sexual desire). A third component (basal ganglia) is responsible for habit formation, movement and elementary learning. These areas of the brain are essential for survival, but the cerebral cortex is of equal, in some respects greater, importance to *human* behaviour.

Notwithstanding the biological complexity just described, evolution, in the context of contractevolism, is about individual organisms having viable offspring, thereby passing on a large portion of their genes to the next generation and beyond. Individuals not driven by a fundamental desire for reproductive success, and the self-preservation required to achieve it,[13] will leave, at best, an extremely faint

---

[13] In some species (arachnids and insects), sexual cannibalism occurs; the female eats her mate. The prevailing evolutionary explanation rests on the benefits and costs for the male. The redback spider is an example. Two factors explain this behaviour. First, males face multiple risks in finding a mate – most suffer predation in the search, for example. The probability of finding a mate before death is low. The probability of finding a second mate is vanishingly small. Hence,

trace in evolutionary history. This *biological* 'self-interest' is what evolutionary history and theory affirm is deeply complex, honed by both selection and culture (Richerson et al., 2003) and includes cooperation. As Frans de Waal (1997) states:

> We are facing the profound paradox that genetic self-advancement at the expense of others – which is the basic thrust of evolution – has given rise to remarkable capacities for caring and sympathy. (p. 5)

There is no paradox but simply a more complex landscape.[14] Evolutionary self-interest is best advanced, in the overwhelming majority of cases, by cooperation and, on a larger scale, sociality. Theoretical and experimental work, as we saw in the previous section, provides a robust explanation of the emergence of reciprocity, cooperation and sociality: caring, sympathy and empathy motivate cooperation.

4.2 Cooperation and Sociality Resulted from Evolution Not Rational Deliberation

Hobbes considered the origin of cooperation, and resultant societies with shared norms, to be a rational response to the state of nature – everyone for herself. By accepting the rules of a society, each individual enhances her or his probability of life-preservation and life-integrity (health). Given what we know about human origins, there was never a time when humans existed in a non-cooperative state of every person for herself or himself. Rather, there is an insensible gradation from non-primate mammals to early primates to human primates, during which propensities for empathy, reciprocity, cooperation and sociality were transformed and honed (Scarre, 2018). At some stage of cognitive development, more successful strategies of cooperation were discovered and passed on through imitation or teaching. Persistence through many generations resulted, in some cases, in the transfer from cultural transmission to genetic propensity, via a process something akin to a Baldwin effect (Deacon, 1997, pp. 322–3; Dennett, 1995, pp. 77–81; Simpson, 1953) or genetic assimilation (Waddington, 1957).

The initial evolution of a propensity for cooperation and empathy occurred long ago in accordance with the dynamics of evolution and largely in the absence of cognition and reasoning. The rise of human societies is driven, in

---

the cost of death is low. There is also a low probability that the female will mate again soon, securing paternity, which is a benefit. The number of offspring is higher and their health greater as a result of the meal, which is a large benefit, although the mechanism is not known (see Andrade, 1996, 2003). In these cases, the desire for self-preservation persists until mating but, obviously, not beyond, demonstrating the primacy of reproduction.

[14] The paradox for de Waal is really a contrast between T. H. Huxley's (1894/1989) 'nature red in tooth and claw' view of evolution and John Dewey's (1898) more nurturing view.

significant part, by this inherited propensity; cognition and reason came later. Hence, a more natural sketch of the dawn of human societies includes individuals who have a genetic propensity to reciprocate and cooperate, have knowledge of their own conditions – strengths and weaknesses, for example – and can hone strategies of cooperation that capitalise on one's strengths and offset one's weaknesses. Trial and error alone is sufficient to discover that offsetting certain vulnerabilities is best achieved by enticing another to assist, but there has to be something in it for that other. That something will almost always require a restraining of some pursuit of one's own interests.

As the evolution of cooperation transitions from an almost exclusively genetic propensity to a mixture of genetic propensity and cultural adaptation, social relations become more complex and reciprocity more nuanced. The emergence of *Homo sapiens* represents the current pinnacle of this process. As a result, modern human societies are more complex than those of other animals, including other primates, and the instantiation in them of 'do unto others as you would have them do unto you' (reciprocity) is more nuanced. Again, there is a gradation, not a saltation, in this increase in social complexity and nuanced interrelationships. The *Homo sapiens* species is currently at the highest rung of the ladder, but *Homo neanderthalensis* and *Homo erectus* are only a rung below and *Homo heidelbergensis* only a rung or two below that (Scarre, 2018). Moreover, the gradation continues from the emergence of *Homo sapiens* to the present (Trigger, 2003).

4.3 Cognition and Propensities

Cognition mediates propensities and there is a spectrum in animal behaviour from unmediated propensities (reflexive: stimulus → behavioural response) to highly mediated ones (stimulus → processing → behavioural response). At one end of the spectrum are sponges and placozoans or, perhaps, the comb jelly species *Mnemiopsis leidyi* (Ryan et al., 2013; Ryan and Chiodin, 2015). Humans are at the other end, although octopod intelligence is also impressive. This spectrum has an understandable correspondence to the spectrum of nervous system complexity. Sponges and placozoans do not have a nervous system, although they do respond to pressure on their surface. Humans are at – or, more cautiously, very close to – the other end. Both ends are fuzzy. Genomic and transcriptomic analyses of sponges, for example, have revealed that they possess a large repertoire of genes associated with neuronal processes in other animals (Leys, 2015), which has led to two hypotheses. The first is that these genes are the precursor to the evolution of a nervous system. The other is that they are 'remnants of a more complex signalling system and sponges have lost cell types, tissues and regionalization to suit their current suspension feeding habit' (Leys, 2015, p. 581; Ryan and Chiodin, 2015). Fuzziness at the other end has to do with complexity. Although the sheer number of neurons is a poor measure of complexity, more sophisticated measures such as the number in regions associated with higher cognitive functioning might possibly place finned pilot whales closer to the current spectral end than humans – they have twice as many neurons as humans in the neocortical region (Sousa et al., 2017, especially pp. 5–6). Sousa suggests:

> One possibility is that increased cognitive abilities emerged following the expansion of the human brain because this expansion untethered large portions of association cortices from previously strong constraints imposed by molecular gradients and neuronal activity patterns, thereby allowing new sets of cortico-cortical synaptic projections, the re-wiring of ancestral circuits, and the development of new behavioral, cognitive, and phenotypic outcomes.
>
> (p. 5; see also Buckner and Krienen, 2013).

This fuzziness does not undermine, however, the general correspondence between the evolution of cognitive-mediation and the evolution of the nervous system, in which the temporal increase in the number, variety and functions of neurons in the cerebrum is critical. What is clear is that the behavioural propensities of sponges are entirely reflexive (non-mediated), and this is true for a large swath of the animal kingdom. It is also clear that, as the nervous system evolves, with it evolves an increasing ability to control evolved propensities. Humans have a sophisticated cognitive capacity, one more than adequate to allow reflection, and modification, of behavioural propensities. The account that Wilson offered in the first chapter of *Sociobiology* was correct in emphasising the powerful behavioural propensities with which evolution has endowed us, the neurological centres of which are the hypothalamus and limbic system, but he was naïve to not place equal emphasis on cognition and deliberation, and their mediating role, which are located in the cortex.

There is a lesson here for contractarians. Rationality of the kind required by social contracts that rely on an original position – or state of nature (Hobbes, Locke, Rawls, Harsanyi) – emerged late in animal evolution. It emerged long after a propensity for reciprocation, cooperation, mutualism and sociality. Although the requisite capacity to reason might predate the emergence of language (40,000–100,000 years ago)[15] (Pettitt, 2018, p. 124), it is nonetheless late in hominid evolution.[16] Hence, insipient social organisation evolved before

---

[15] The FOXP$_2$ gene (a gene on chromosome 7) is a single mutation gene that is now widely believed to facilitate language. It appeared no more than 100,000 years ago.

[16] Hominids emerged about 6–8 million years ago (*Sahelanthropus tchadensis* and *Orrorin tugenesis*) (Toth and Schick, 2018, p. 57). Language emerged no earlier than 100,000 years ago

the possibility of rational-choice. Skyrms is correct to reject contractarianism and focus on the *evolution* of a social contract. This involves the evolution of both the propensities underlying it and its specific early forms. Game theory can explain how cooperation is selectively advantageous, but a complete picture must connect this mathematical framework to population and molecular genetics. Behavioural propensities with which evolution has endowed us, and the social interactions to which they have led, are temporally prior to rational deliberation.

Prior to the dawn of agriculture, groups were small. Whether agriculture enabled increases in population density or population density forced agriculture and domestication (Josephson et al., 2014; Kavanagh et al., 2018), the two are historically linked. Hence, by the time larger societies emerged (10,000–15,000 years ago), the dynamics of social groups included rationality, which emerged about 100,000 years ago and culture, which emerged much earlier.[17] The development of complex societies and the moral and social codes within their fabric was undergirded by evolved propensities but clothed by rational decisions and cultural evolution.

4.4 Inclusivity, Liberty and Equality

Reciprocation, cooperation and social organisation have evolved because those behaviours have enhanced the preservation of the ERS of the individuals manifesting them, where ERS includes the preservation of life and health. Evolutionary dynamics explains this (Alexander, 1974, 1987; Axelrod, 1984; Hamilton, 1964a, 1964b; Hammerstein, 2003; McElreath and Boyd, 2007; Trivers 1971). It is the enhancement of the ERS of individuals that drives and *justifies* the fabric of a social organisation. *It is here that description becomes prescription.* The propensity for cooperation and the resulting social organisation are the product of evolutionary dynamics. Cooperation and social organisation are the evolutionary solution, wrought over millions of years, to the challenge of maximising the fundamental interest of individual organisms: ERS. That solution has bound together the fate of individual *Homo sapiens* and depends necessarily on solidarity, that is,

---

(Pettitt, 2018, p. 124), during the last 7 per cent of hominid evolution. The earliest known use of tools is 2.5 million years ago (Toth and Schick, 2018). Making and using tools was a significant evolutionary transformation. Tool construction and use require some level of reasoning. Hence, it is possible that some – minimal, it would appear – cognition was brought to bear on cooperative strategies at this point, that is, in the latter third of hominid evolution, long after the propensities crucial to society-forming (obviously cooperation and empathy but also territoriality, as well as mating and nurturing propensities) had been formed in non-human primates and also animals lower on the evolutionary tree.

[17] See Note 15.

on an *obligation* on each person to maximise the *opportunities* for enhancing the ERS of themselves and others. Moral norms, rules of social organisation and conventional norms of behaviour are the fabric of a society and are the social manifestation of the obligation. Morality is a social phenomenon. On this view, morality can only exist inside a social contract. A social contract that maximises the *opportunities* for the ERS of all members maximises the 'Good'. Maximising opportunities, however, does not entail requiring that one take advantage of those opportunities. Those that do not take them risk the evolutionary extinction of their particular genotype but, for cognitive organisms, that choice is available. What is immoral (evil) is reducing these opportunities for other individuals within the contract (Thompson, 2002). This perspective underscores that, contrary to Hobbes, this obligation is not an edict or product of reason, or at least not fundamentally and initially so; it is a consequence of evolutionary dynamics.

Although humans have been successful in harnessing the dynamics of nature – from building reservoirs to nuclear-generated electricity to modifying the female ovulatory hormone cycle – failure to respect the fundamental dynamics of nature still spells doom; pandemics, antibiotic-resistant bacteria and climate change are stark reminders. Consequently, as a *reductio ad absurdum*, any moral theory that fails to promote individual ERS is a path to extinction. One can modify or supress any specific behavioural biological propensity; that is how the bonds of biological determinism are broken. However, if the total collection of one's *chosen* behaviours negatively affects that person's ERS, that person's lineage is doomed; hence, if those behaviours are morally required within a society, the society is doomed. This is why the fundamental moral 'Good' is the promotion of ERS.

This fundamental 'Good' justifies the moral norms of a stable, functioning society. Membership in a society results in the fate of each individual being significantly tied to the common fate of all, which is itself dependent on the stability and continued existence of the society. Consequently, the proximate justification of a set of moral norms (albeit a changing set over time) depends, in large measure, on demonstrating that the social fabric is stable (resistant to disintegration and invasion) and well-functioning. 'Well-functioning' means that, taken as a whole, the moral norms of the society secure the maximal promotion of the ERS of its members.

Of course, the fabric of the society will include many more norms and practices than moral norms. The warp and weft of the social fabric include its laws, judiciary, political economy and engagement of citizens. Prima facie marks of trouble, and hence the need for social change, will be social instability and functional inefficacy. Although social stability mirrors in some respects the 'evolutionary stable strategies' of Price and Maynard Smith, there are important differences. An evolutionary stable strategy is a behavioural strategy, which, when adopted by all members of a population (or a high percentage), is resistant to an alternative strategy increasing in frequency. Social stability is more multifaceted.

The foundational 'Good' permits a presumptive justification of three fundamental, high-level moral norms: inclusivity, liberty and equality. The justification rests on the dynamics of the interaction of humans within an environment that constrains the ERS of their behavioural options. As such, it is a contextual, empirically based justification, not a logical derivation.

It has become common to express these norms in the language of 'rights'. This is regrettable because the emergence, since the eighteenth century, of a plethora of rights has led to confusion. As Wayne Sumner (1989) has noted:

> If we once pause to reflect on the bewildering array of rights invoked in both domestic and international affairs we cannot avoid asking ourselves hard questions. Which of these rights are genuine and which are not? Which deserve to be taken seriously and which do not? And in cases in which there is a genuine right on each side of the issue, which deserves to be taken more seriously? (p. 4)

Here, I avoid the language of rights.

Simplifying slightly, humans face three major challenges to their ERS: non-human predators and aggressors, human aggressors, and the vagaries of nature (e.g., flood, drought, famine, earthquakes, volcano, disease). Physical strength increases the probability that an individual will be successful in fending off human and non-human aggressors. Locomotive speed increases the probability that an individual can escape a predator or aggressor. Mental prowess (fast-learning, speed of information-processing, problem-solving ability, for example) and acquired knowledge will increase the probability of survival in a broad range of contexts, including the threat from disease; being able to quickly discern the early symptoms of a disease and how it spreads enhances avoidance of high-risk situations. Knowledge via experience and education enhances avoidance of danger and thereby enhances the 'Good'. The examples of characteristics useful in particular circumstances can be multiplied easily; these will suffice here.

In a group, small or large, some individuals will be able to contribute less (those who are weak, physically and mentally slow or perhaps infirm, for example) and others more (those who are physically strong or mentally quick, for example). Given this disparity, why should anyone care much about those who have little to contribute? Why should those with a wealth of advantages even participate in cooperative arrangements? The simple answer is because individuals are not isolated entities.

The poorly endowed have parents, siblings, aunts, uncles and cousins, and we know that kin cooperation enhances the ERS of kin members, which has resulted in strong bonds of kin cooperation. In small societies, a kin group can collectively bring grief to the ERS of others; that applies even if each member of the kin collective were poorly endowed; united they stand. More importantly, however, the poorly endowed can form other alliances with others who are poorly endowed. They can be local reciprocators, compensating for each other's weaknesses. There will be very few who cannot enhance their bargaining position through local solidarity with kin and others.

The well-endowed are not isolated either. They will have relatives on whom their inclusive fitness will depend. Hence, reciprocation matters to them as well; how they will fare against others who are well-endowed is uncertain and alliances are prudent. There will be very few for whom 'going it alone' will be the best strategy for maximising their fundamental 'Good'. Moreover, attempts to exclude any individuals will be risky and will consume time, resources and attention (watching one's back, for example). Hence, excluding anyone from the fold is a high-risk strategy.

Moreover, the future is unpredictable; by misfortune, failure or changes in the environment, some of the well-endowed will tumble into the group of less well-endowed, and on occasion some of the less well-endowed, by good fortune or changes in the environment, will enter the group of well-endowed. Changes in the environment can result in characteristics that were once strengths becoming less so or even liabilities and vice versa. Everyone will have witnessed, to some extent, these vagaries of life; life is uncertain and unpredictable. There is no need for a veil of ignorance for individuals to be motivated to be inclusive of both the well-endowed and the poorly endowed and everyone in between. Human experience of the dangers of exclusion and the unpredictability of life will be motivation enough. Human experience, however, is piggybacking on hundreds of thousands of years of natural selection, during which a propensity for 'going it alone' has been selectively disadvantageous for the same reasons.

Maximal liberty (freedom of choice) is also derivable from the 'Good'; harnessing cooperation in the service of protecting ERS requires free agents. Servitude is not cooperation. The foundation of contractevolism is that there is an evolutionary imperative for an individual to maximise the *opportunities* for her own ERS; the trial and error of evolution over hundreds of millennia has converged on the principle that, to maximise the *opportunities* for one's self, one must cooperate. This generates an obligation to maximise equitably the *opportunities* for the ERS for all members of her group. This is a hollow obligation if the person is not free to discharge it, even though discharging it within a society will require every member relinquishing some freedoms.[18] To be coerced to relinquish more exceeds the requirements of maximising the 'Good' for all members. Consequently, a coercer has broken the reciprocal/cooperative agreement. To be able to fulfil an obligation, one must be free to choose, that is, to be in control of one's destiny consistent with that same freedom for others. A person cannot 'do unto others as she would have them do her' unless she is free to 'do unto others'.

A social order in which one person, or even a group, can coerce the behaviour of some others in order to enhance their own ERS at the expense of the ERS of the others is, tautologically, not one that maximises the ERS of all engaged in the venture. Reductions in individual liberty, consequently, must be the minimum necessary to achieve the benefit. Greater encroachment on an individual's liberty by others and/or the society risks individual defection, which is the seed of insurrection and revolution – at a minimum instability. I find the concepts of individual sovereignty and independence as developed by Arthur Ripstein (2004, 2006) to be a compelling expansion and exploration of the implications of cooperation-based liberty:

> The sovereignty principle's focus on voluntary cooperation also explains why other harms fall outside its scope. Voluntary cooperation enables people to use their powers together to pursue purposes they share. It can be made to look as though potential cooperators are always subject to each other's choice: unless you agree to cooperate with me, I cannot use my powers in the way I want to. But this is an example of our respective independence. Cooperation contrasts with domination only when it is voluntary on both sides. You get to decide whether to cooperate with me because you get to decide how your powers will be used. I can no more demand that you make your powers available to accommodate my preferred use of my powers than you can make that demand of me. Each of us is sovereign over our powers, and the power to decide with whom to cooperate is a basic expression of that sovereignty. That is why I wrong

---

[18] Some might associate the views here with those of Herbert Spencer (1851) in his *Social Statics*: 'liberty of each, limited by the like liberty of all, is the rule in conformity with which society must be organized' (p. 88). Spencer did provide a primitive evolutionary basis for his utilitarianism and his connection of justice and liberty. *Social Statics* was published in 1851, prior to Darwin's *Origin*. While I agree with his conception of liberty, the evolutionary basis given here has the benefit of 170 years of evolutionary theorising and advancement of knowledge, and of philosophical analysis and reflection. Moreover, it is social contract theory not utilitarianism that underpins it.

> you when I use your powers for my purposes, even if it does not cost you anything: in appropriating your powers as my own, I force you to cooperate with me. (Ripstein, 2006, p. 237)

Although here I only offer a promissory note, I hold that, on contractevolism, justice is liberty. A just society is one in which freedom to pursue one's own interests is unimpeded by others beyond what is required to maximise the equivalent freedom of all members.

Equity is the third fundamental principle derivable from the contractevolist 'good', although it might also be regarded as a corollary of liberty. Recent modelling of the evolutionary dynamics of egalitarianism as well as related experimental work suggests that there is an evolutionary propensity for egalitarianism (Bolton and Ockenfels, 2000; Calmettes and Weiss, 2017; Dawes et al., 2007; Fehr and Schmidt, 1999; Gavrilets, 2012b; Rabin, 1993). As such, it provides support for the view that evolutionary dynamics have shaped yet another important feature of our behavioural repertoire. Since these behavioural propensities have evolved to maximise individual ERS, they support the derivation of equity of treatment from the 'good', which is itself grounded in maximising ERS.

There are lots of impediments to equity. Those of different abilities will have different options open to them: different opportunities. Different circumstance also will affect the options available. Like the simple principle of determinism (same cause, same effect), the simple concept of equality (treating like individuals in like circumstances in the same way) is empirically illusory. Two causal situations are almost never actually identical, and two individuals and their circumstances are almost never identical. High levels of abstraction and simplification are required to create the identity – sometimes productively, often not. The best one can achieve in a social contract is a social fabric in which individuals will have maximal freedom to pursue their own interests and the mitigation of impediments and unequal possibilities for achieving them.

The mathematical lesson of long-term (indefinitely repeated) games is that a social structure can mitigate some inherent natural impediments; working together, equity can be enhanced for all (Binmore, 1994, especially pp. 114–25). As Binmore insightfully remarks:

> I am delighted to be able to quote Confucius as authority on this point. When asked if he could encapsulate the 'true' way in a single word, he replied: 'reciprocity'. (p. 114)

As Binmore notes, translations differ but all express the primacy of reciprocity. For example, the Chinese ideogram *shu* is sometime translated as 'the not doing to others what one does not like oneself', which is a variant on the golden rule and reciprocity.

4.5 Resolving Sumner's Social Contract Challenge

As noted, maximisation of equity within a society also maximises liberty and vice versa; hence, both liberty and equality rise and fall together, tied to the fate of the ability to achieve one's interests – fundamentally ERS. Herein lies the contractevolist answer to what I shall call *the Sumner challenge* (after Wayne Sumner) to the social contract theories of Hobbes and Gautier:

> One of the functions of a set of moral rights is to negate the advantages of predation by providing equal protection for all. But equal protection will not be the outcome of a rational bargain if the strong and ruthless are allowed to bring their ill-gotten gains to the bargaining table. (Sumner, 1989, pp. 157–8)

> Gautier may be right in thinking that rational individuals in the state of nature will manage something better than Hobbes' war of all against all. But as long as instrumental rationality permits some exploitation by the strong of the weak then natural interaction will fail to be a fair initial situation for agreement on a set of rights. (Sumner, 1989, p. 162)

Rawls' social contract does better than Hobbes'. Rawls' key moral concept is fairness; the veil of ignorance is supposed to achieve this and hence mitigate Hobbes' unlevel starting point, but this simply moves the focus to 'how are these antecedent standards of fairness to be justified?' (Sumner, 1989, p. 159).

Unlike contrarianism, contractevolism is not a contract constructed by rationally deliberating agents starting with an individualistic state of nature or original position. It is evolutionary. Evolutionary dynamics work against the 'strong and ruthless', promoting instead genes for a propensity to reciprocate and cooperate. This entailment of the dynamical principles is entirely consistent with the empirical evidence; cooperation and reciprocity are pervasive. The evolution of cooperation is, in a significant way, the evolutionary answer to the domination of the strong and the ruthless. One need not be naïve in order to place emphasis on the evolution of cooperation; one can accept that competition (especially for mates) and territoriality are also pervasive and important, a point to which I return in the next section. Nonetheless, cooperation has evolved as a balancing influence on the domination of raw power and 'ill-gotten gains'.

That said, history is strewn with examples of group hegemony and jostling for advantage. Robert Skidelsky (2020), in the context of contemporary economics, has crisply summarised this:

This chapter is directed at the neglect by economists of the role power plays in economic relations. This neglect is deliberate. By ignoring the extent to which power pervades the economy, mainstream economists buttress existing structures of power by rendering them invisible. (pp. 119–20)

For contractevolism, Sumner's requirement of a fair starting point, or a level playing field, is otiose; social groupings, from kin groups to small tribes to large complex societies, are temporal segments of evolutionary history. As already observed, there never was a moment in time when rational deliberation resulted in a social contract. There is a gradual transition during animal evolution in which the dynamics of evolution promote advantageous behaviours, and cooperation is a powerful advantageous behaviour. At some point, reason and cultural transmission of learned strategies became part of the process but the demands of ERS relentlessly sift the behavioural repertoire.

There is nothing static about this picture. When features of nature change (climate, food resources, novel predators, for example) or social arrangements change (invasions, insurrection, hegemonic transitions, increased suppression of liberties, widening economic or other inequalities, for example), the dance begins again until a new equilibrium is found. *Homo sapiens*, more than other species, transform both their behaviour and the world in which they live. Ultimately, however, even *Homo sapiens* dance to the tune of the evolutionary and physical dynamics of life and the universe. They can change the environment to their 'supposed' benefit but new behaviours that work against the maximisation of the ERS of individuals will be eliminated or, if the change is too dramatic, the saga will end in misery – at the limit, the end of the species. Climate change might be such an end.

Inclusivity, liberty and equity enhance the achievement of the 'Good' – the maximisation of the ERS of individuals within a society. I hold that it is not a coincidence that other moral theories encompass inclusion, liberty and equity. The explanation is that our biological propensities to reciprocate, cooperate, empathise and socialise are cognised as moral sentiments and these sentiments entail inclusivity, liberty and equality. In that sense, these truths are self-evident. Wilson was correct that the sentiments 'consulted by moral philosophers' are the product of evolution, but he ended the chain of reasoning too close to its beginning.

4.6 A Dismal Reality

Notwithstanding all the above, there is a dismal reality to be faced. Evolutionary dynamics explain the presence and pervasiveness of reciprocity, cooperation and sociality. In cognitively advanced organisms, such as humans, culture also has shaped and honed qualities that enhance ERS and these have been transmitted through imitation, inculcation and learning. The apparent dismal reality is that other propensities also enhance ERS: territoriality, suspicion of outsiders, retaliation for grievances, greed and mate possessiveness, to mention but a few. What has evolved is a complex intertwined collection of propensities that are in a stabilising creative tension – analogous to the physical and chemical forces that in a stabilising creative tension hold our solar system together and the objects therein. Viewed this way, the dismal picture fades slightly; creative social tension is an unavoidable property of the fabric of a society. Measured behavioural expressions of territoriality, xenophobia, retaliation and the like have evolved because they enhanced ERS. Territoriality, for instance, although beneficial to individuals, is beneficial to a cooperating group. It creates solidarity within and inhibits encroachment from outside individuals or groups. Untamed, however, these propensities have been exceptionally destructive. They are in part tamed by cooperation, by an expectation of reciprocity. Outsiders who demonstrate reciprocity can become insiders (friends); insiders who violate expectations of reciprocity become outsiders (enemies) and are severed in some way from the group. Morality emerges as an essential component in the system.

Morality across societies differs. Different circumstances justify different specific moral norms. What contractevolism requires is that moral norms across all societies maximise ERS and this requirement entails inclusivity, as well as maximal liberty and equity. Differences in moral codes over time and geography result more from the requirement that the society functions well and is stable. Nonetheless, a stable functioning society must also craft a moral code and social norms that are inclusive and maximise – recognising all the trade-offs – the liberty and equity of each citizen.

Moral codes can differ over time and geography because they fall short of achieving the goals of moral codes in different ways and for different reasons. This simply reflects the creative tensions pulling and pushing in different directions, with a rebalancing here and a new disturbance there. All currently espoused moral systems (contractarian, contractualist, utilitarian, Kantian) struggle with the same problem; social systems fall short of the demands of the moral theory. Contractevolism has an explanation and a prescription; moral progress occurs when cognitive strategies for enhancing ERS in social contexts overrides evolved propensities that no longer enhance ERS within groups of cooperating cognitive individuals. Successfully overriding these propensities requires robust knowledge of the origins and manifestations of them; evolutionary theory, evolutionary history and biological/anthropological/psychological knowledge are, hence, crucial.

4.7 Free-Riders and Miscreants

There is a further dismal (at least for cooperators) reality: free-riders and miscreants. The major operational difference is that a free-rider takes the benefits of the cooperative actions of others without being a contributor – that is, by not cooperating – or, more generally, obtaining a benefit that others make possible without participating in the production of the benefit. By contrast, miscreants (villains) do not exploit the good will of others; they *break* the social rules. Psychopaths and rapists, for example, are not free-riders; they reject the social fabric. Miscreants, as thus defined, destabilise the society unless removed or rehabilitated (brought back into the cooperative venture), and societies typically devise methods of doing precisely that, with greater and lesser degrees of success, but the goal is clear. Destabilising the society is the paradigm of evil for contractevolism (Thompson, 2002). Free-riders, however, constitute a different challenge.

Free-riding is only evolutionarily successful when it coexists at a low frequency in the population – low enough that it eludes attention. In effect, there is frequency-dependent selection. From an evolutionary perspective, free-riders increase their ERS by exploiting the efforts of others without negative consequences. As the frequency of free-riders increases, however, the probability of consequences delivered by the generators of the benefit increases. Below some system-relative threshold frequency, usually very low, either free-riders will be undetected or attempts to supress them are counterproductive. At or below that threshold frequency, free-riders have found a stable strategy of exploitation of the cooperative behaviour of the society; further invasion of the dominant strategy will destabilise the entire enterprise. As long as further invasion renders the free-rider strategy selectively disadvantageous, the society of cooperators is a stable strategy. The two strategies coexist. Far from undermining a moral system based on enhancing the opportunities for ERS through cooperation (and through the principles of inclusion, liberty and equality), the existence of a low threshold frequency – beyond which further invasion is selectively disadvantageous – demonstrates the robustness of the cooperative, morality-generating strategy. Free-riders differ from other miscreants by depending on the stability and survival of the majority: the cooperating individuals. Psychopaths do not have this relationship to the majority.

5 Hume's Barrier and Moore's Fallacy

5.1 Hume and Naturalism

At the end of book 3 ('Of Morals'), part 1, section 1 of David Hume's *A Treatise of Human Nature*, he wrote:

I cannot forbear adding to these reasonings an observation which may, perhaps, be found of some importance. In every system of morality, which I have hitherto met with, I have always remark'd, that the author proceeds for some time in the ordinary way of reasoning, and establishes the being of a God, or makes observations concerning human affairs, when of a sudden I am surpriz'd to find, that instead of the usual copulations of propositions, is and is not, I meet with no proposition that is not connected with an ought, or an ought not. This change is imperceptible, but is, however, of the last consequence. For as this ought, or ought not, expresses some new relation or affirmation 'tis necessary that it shou'd be observ'd and explain'd, and at the same time that a reason should be given for what seems altogether inconceivable, how this new relation can be a deduction from others, which are entirely different from it. But as authors do not commonly use this precaution, I shall presume to recommend it to the readers; and am persuaded, that this small attention wou'd subvert all the vulgar system of morality, and let us see, that the distinction of vice and virtue is not founded on the relations of objects, nor is perceived by reason. (Hume 1738/1960, pp. 467–70)

This section, 'Moral Distinctions Not Derived from Reason', is devoted to his rejection of the rationalist view of morality. Three main factions existed when Hume was writing: Hobbesian (self-interest/selfish basis of morality), rationalism and sentimentalism – with the latter of which Hume identified. In the *Treatise*, Hume appears to simply assume without argument that the Hobbesian view is untenable; he sets out his reasons for rejecting it in his later *Enquiry* but even there in an appendix (Hume 1748/1963).

In the almost 300 years since Hume wrote the above-cited passage, the interpretations of it have been numerous and inconsistent. The view that has come to pervade contemporary philosophy is that it is a *logical* fallacy to support a moral (ought) conclusion using factual (is) premises alone. All such arguments (usually structured as syllogisms) are either irreparably fallacious (invalid) or enthymemes (arguments with assumed premises).

If Hume's comment was merely an observation about the nature of deductive logic, it is obviously correct but trivial. Hume's observation, however, is part of his larger project. It is part of his case against moral rationalism and his defence of his own moral sentimentalism. For Hume, moral rationalism is committed to 'oughts' being discovered by reason, a priori or from the nature of things. His is/ought distinction lays bare that moral precepts must precede reasoning. Hume's sentimentalism is firmly rooted in the nature of things, so he cannot be taken to hold that moral precepts cannot be *justified* by the nature of things. His position is that a *logical* justification is the wrong stance. His sentimentalism is empirical in nature and, contrary to the interpretation of some, is naturalistic. His logical observation undermines rationalism; moral claims cannot be discovered or justified by logical deductions. They can be unpacked through reason but the fundamental 'oughts' rest on a different process, namely attention to sentiments. Charles Pigeon (2019) deftly dispatches the anti-naturalist conclusions drawn by many from Hume's claim:

> MacIntyre and Hunter argue that since Hume was a naturalist in ethics he did not believe that you cannot derive an *ought* from an *is*. Atkinson and Flew reply that since Hume believed that you cannot derive an *ought* from an *is* he was not a naturalist in ethics but must have been a forerunner of non-cognitivism. Both parties are dependent on the same basic mistake. For they both presuppose that *if* Hume was denying the possibility of Is/Ought 'deductions' he must have been denying the possibility of deriving moral conclusions from non-moral premises with the aid of analytic bridge principles. But if Hume was only a *Logical* Autonomist – that is, if he was only denying the possibility of *logical* deductions from non-moral premises to moral conclusions – then there is no contradiction between his meta-ethical naturalism and No-Ought-From-Is. And the fact is that Hume was only a Logical Autonomist. (p. 93)

Pigeon concludes his chapter with, 'These, then, are the mistakes and that is their history. It's a sad and sorry tale. Let's try to do better in future' (p. 95). Hume's is/ought distinction has been wrenched from the context of his moral philosophy and his logical autonomism.

The foundation of Hume's moral theory is sentimentalism. Of the three moral foundations current in Hume's day – selfish nature (e.g., Hobbes), rationalism (e.g., Locke, Samuel Clarke and William Wollaston) and sentimentalism (e.g., Anthony Ashley Cooper (Earl of Shaftsbury) and Francis Hutcheson) – Hume defended sentimentalism. Hume's moral sentimentalism is complex, but in essence he holds that virtue and vice produce in us emotions – love and pride and hatred and humility, respectively:

The chief spring or actuating principle of the human mind is pleasure or pain . . .

> Now since every quality in ourselves or others which gives pleasure, always causes pride or love, as every one that produces uneasiness excites humility or hatred, it follows that these two particulars are to be considered as equivalent, with regard to our mental qualities, *virtue* and the power of producing love and pride, *vice* and the power of producing humility or hatred. In every case, therefore, we must judge of the one by the other, and may pronounce any *quality* of the mind virtuous which causes love or pride, and any one vicious which causes hatred or humility. (Hume 1738/1960, pp. 574–5)

Underpinning much of his sentimentalism is the working of sympathy (closer to empathy in today's use):

It is worth noting that, on their [Hume's and Adam Smith's] shared view, sympathy plays two different roles. First, sympathy with the plight of others engages our concern and prompts our actions in ways that are, they hold, morally important, crucial for constituting and sustaining a community, and more generally mutually advantageous. Second, sympathy is essential, as they see it, to our capacity to approve (or disapprove) of actions, motives, and characters as moral or not and, because of that, to our capacity to judge actions, motives, and characters as moral or not. Thus, without sympathy we would not have a morally decent community, if we had a community at all (that is sympathy's first role), nor would we be able to judge communities (or anything else) as morally decent or not (that is sympathy's second role).

(Sayre-McCord, 2013, p. 209)

It is easy to see parallels between Hume's sentiments (love, pride, hatred, etc.) and E. O. Wilson's 'emotional control centres in the hypothalamus and limbic system of the brain', which 'flood our consciousness with all the emotions – hate, love, guilt, fear, and others'. There are some obvious differences but there is nonetheless a link.

5.2 Moore and Non-naturalism

G. E. Moore was explicitly a non-naturalist. For him, moral claims are true or false in some objective sense – although there is no moral 'reality' akin to empirical reality – but they are not derivable from (or reducible to) non-moral claims (scientific or metaphysical claims). He held that the justification of moral claims rests on intuition; they are claims that are self-evident.

Moore crystallised his position that moral claims are not the same as – not derivable from nor reducible to – scientific or metaphysical claims by labelling any attempt to do so as a fallacy: the naturalistic fallacy:[19]

It may be true that all things which are good are *also* something else, just as it is true that all things which are yellow produce a certain kind of vibration in the light. And it is a fact, that ethics aims at discovering what are those other properties belonging to all things that are good. But far too many philosophers have thought that when they named those properties they were actually defining good; that these properties, in fact were not 'other' but absolutely and entirely the same with goodness. This view I propose to call the 'naturalistic fallacy' and of it I shall now endeavour to dispose.

(Moore, 1903, p. 10)

This is the fallacy that a moral naturalist *must* commit, since good, like yellow, corresponds to certain natural properties but is not what we actually experience.

---

[19] Baldwin (1992) has identified three versions of the naturalistic fallacy in *Principia Ethica*. Here, I quote the most commonly cited statement. The other versions highlight different aspects of his claim that good is indefinable, non-natural and not equivalent to anything else.

Yellow, as a perception, is not identical to a certain wavelength of light nor the effect of the wavelength on the retina of the eye nor the neurological activity created by the activity of the retina. Yellow is a different kind of property from the concatenation of the natural properties with which it corresponds. Similarly, good is a different kind of property than natural properties. To hold otherwise is to commit a fallacy of reasoning.

If good is to be understood in the same way as yellow (non-natural, non-material), then both are mental, in some sense, and not physical properties, as is pain. Good for Moore, however, although analogous to yellow, is different from yellow and also from pain. Yellow and pain are caused by (or, to use Moore's terminology, 'correspond to') actual physical things or properties; good apparently does not.

> We might think just as clearly and correctly about a horse, if we thought of all its parts and their arrangement instead of thinking of the whole, we could, I say, think how a horse differed from a donkey just as well, just as truly, in this way, as now we do, only not so easily; but there is nothing whatsoever which we could substitute for good; and that is what I mean, when I say that good is indefinable. (Moore, 1903, p. 8)

Because 'fallacy' in logic refers to an error in reasoning, the use of it here might suggest that Moore has discovered/revealed an invalid form of proof employed by naturalists but that is not the case; he has simply asserted that it is an error to define good in terms of natural properties. To the extent that he offers a 'proof', it has been found by many in what has become known as his open question argument – a label given to the argument by others.

Moore has as his target Herbert Spencer's evolutionary ethical views and Bentham's utilitarianism. The simple principle of utilitarianism will suffice here: one should act so as to bring about the greatest good for the greatest number. Good is most broadly understood as the satisfaction of interests but, for Bentham, and others, good was happiness (or pleasure). Assuming that the claim 'α is good' is logically equivalent to 'α is pleasure', the claim that pleasure is good reduces to the tautology 'pleasure is pleasure' or 'good is good'. This is one version of the open question argument. This strikes me, as it has many others since 1903, as simplistic. Any definition of a thing in terms of its properties is ultimately tautological. To use a well-worn example, if 'β is water' is logically equivalent to 'β is H_2O', then the claim water is H_2O reduces to the tautology H_2O is H_2O.

A more substantive rendering of Moore's open question notes that, for any naturalistic claim about good, one can always ask 'but is it good?' Consider a situation in which it is uncontroversial that person A is experiencing pleasure.

If pleasure is good, the question 'but is it good that A is experiencing pleasure?' is superficially meaningless. This, to Moore and those who find his open question argument compelling, appears implausible. Consequently, good cannot be identical, or reducible, to pleasure. This holds true of any other natural property used to define good. If good is financial success, one can ask, meaningfully according to Moore, 'is it good that she is financially successful?' The assumption – an assumption that makes it a meaningful question – is that it could be answered negatively. In which case, good cannot be defined by financial success or reducible to it.

From the vantage of the early twenty-first century, this all seems a bit quaint. The logical empiricists discovered that definitions of empirical things and properties were multifaceted. Correspondence rules (bilateral reduction sentences) gave partial definitions and, hence, where relevant, partial reductions. Indeed, exhaustive definitions were usually unfeasible. Tarski's model-theoretic semantics provided a much better route for interpreting terms than dictionary-style definitions. But that requires a holistic view of a domain of discourse, which Moore's approach is not. The term 'good' and its meaning(s) are part of an interconnected network of discourse. Even staying with definitions in terms of *A* is *B*, Wittgenstein's family resemblance captures well the complexities and pitfalls of separating a word from its place in a domain of discourse.

As the dust has settled since 1903, the problems with Moore's anti-naturalism have become clear. To be fair, the exploration of Moore's positions and arguments have also been fruitful in ways well beyond the simple determination of the plausibility of his claims and arguments. Neil Sinclair's (2019) edited volume illustrates this well.

The conclusion of Nicholas Sturgeon's (2003) article in *Ethics* captures correctly where a century of pondering Moore's *Principia Ethica* has landed us:

> My reason for calling attention to these further complications, and further arguments, is not, however, that I think that Moore's arguments succeed. What I have argued, in fact, is that the familiar arguments face even more difficulties than are usually acknowledged, and that the unfamiliar ones fare little better. So I do not think that Moore mounted a damaging case against ethical naturalism. (pp. 555–6)

There is another perspective on Moore's moral framework. His non-naturalist stance led him to intuition as the basis for understanding 'Good'. To the extent, however, that our current sentiments and intuition are part of our 'mental' repertoire, they are there because they promoted the 'Good', as I have portrayed it (that is, as embedded in evolutionary dynamics), namely ERS. Obviously,

I exploit that fact in drawing out inclusivity, liberty and equality from the fundamental Good. It should surprise no one that such key principles seem intuitively correct. If I am correct, the moral concepts and the fact that they appear intuitive – to the extent they are –both derive from the dynamics of evolution.

A final word: what appears to be at stake for Moore and many twentieth-century philosophers is normativity, which stands in for a concern for the independence of ethics. I am unclear on what role independence plays here but suspect that the discomfort is that accepting a naturalistic moral theory means that morality, ethics and metaethics fall in some manner into the domain of empirical science. It is not clear that this conclusion is correct. After all, utilitarianism is naturalistic but interpreting it and exploring its implication for behaviour have, for nearly two centuries, remained within the subfield of philosophy called 'ethics'. Equally, social contract theory has remained embedded in ethics and political theory, just as it was in Hobbes' writings. No doubt, a contactevolist naturalistic theory will require greater attention to evolutionary theory, and to biological and psychological work on human behaviour, and perhaps animal behaviour more generally, but the autonomy of ethics need not be abandoned. Normativity is not at stake; norms of behaviour will always be an important subject as long as humans share an environment and the complex structure of that environment continues to change.

6 Contractevolism and Patriarchy

In Sections 3 and 4, I explicated the essence of contractevolism and in Section 5 dispelled two long-standing objections to such naturalistic views of morality. I now turn to its implications for several social issues: patriarchy, copulatory behaviour and overpopulation and extinction.

This section explores an apparent tension within contractevolism, which claims to derive the principle of equality from the moral primacy of maximally enhancing the opportunities for ERS. It does so in the face of undeniably pervasive, although highly variable, evolved male coercive mating behaviours that enhance the ERS of males. In the human context, a manifestation of these mating behaviours is patriarchy; males rig the social organisation to advance their ERS, along with many collateral social advantages.

6.1 Religion and Patriarchy

Male domination of women through social rules and practices, and through control of the mechanisms of social power, has been pervasive, though not universal, throughout human history. Although the focus here is on biological evolution and sexual equality, there is a cultural dimension that, while relevant to ERS, constitutes a much broader framework for sexual inequality, one rooted in primordial ideologies. Various religions codify these ideologies and, in many societies, have been instruments of patriarchy. Jainism and Hinduism appear to encode patriarchal ideologies the least; Abrahamic faiths (Judaism, Christianity and Islam) the most. For example, Christianity has underpinned an almost two-millennia history of patriarchy in European and English-speaking countries. In the very early days of Christianity, there is evidence that women played central roles. The institutionalisation, legalisation and politicisation of Christianity changed all that. Judith M. Leiu (2013) provides an excellent examination of the changing role of women in the early church as men came to dominate both its governance and the codification of the 'faith' through councils. It is a depressing history. Augustine of Hippo, whose ideas continue to influence Roman Catholic theology and beyond it to other sects of Christianity, saw no meaningful free will in humans, only a soiled and destroyed nature – a consequence of the fall from grace of Adam and Eve:

> His [Augustine's] full answer involved doctrines of total depravity, arbitrary election, predestination, the physical inheritance of the original taint through man's sexual nature, and the damnation of the unbaptized.
>
> (Ferguson, 1980, p. 119)

> Augustine comes to insist on the far more radical notion that God loves in his children only the virtues that he implants there himself: even the very desire for a virtue is a divine gift. (Wetzel, 2012, p. 340)

The concept of the 'inheritance of the original taint through man's sexual nature' plays a significant role in Augustinian misogyny and patriarchy. Pelagius, by contrast, espoused human free will and culpability for wrong actions, as well as human moral frailty, but not inherent depravity. For political and economic reasons much more that scriptural ones, Augustine's position prevailed (see Beck, 2007); Pelagius was deemed a heretic at the Council of Ephesus in 431. However,

> If a heretic is one who emphasizes one truth to the exclusion of others, it would at any rate appear that [Pelagius] was no more a heretic than Augustine. His fault was in exaggerated emphasis, but in the final form his philosophy took, after necessary and proper modifications as a result of criticism, it is not certain that any statement of his is totally irreconcilable with the Christian faith or indefensible in terms of the New Testament. It is by no means so clear that the same may be said of Augustine.
>
> (Ferguson, 1956, p. 182)

Christian history would have been very different if Pelagius had been the victor. What impact his views might have had on Christian patriarchy is unknowable. Since the role of religious ideology in the history of patriarchy, however, is not the focus here, and this is not a treatise on theology or church history, we must move on.

6.2 'On Average' Sexual Dimorphisms

Sexual dimorphism is undeniable. Some differences are visually obvious (breasts and penises, for example); others, such as hormonal cycles, are not. Does sexual dimorphism justify sexual inequalities, specifically some form of paternalism, on a contractevolist theory? I begin by clearing away some rubble. One class of evolved differences between women and men poses no challenge to contractevolism's equality principle. Height is an example; women *on average* are shorter than men, but both vary on a spectrum. The variation conforms to a Gaussian (normal, bell) curve distribution. Data is not available for all countries, but based on twenty countries, which vary geographically and socio-politically (Jelenkovic et al., 2016), the distribution in Figure 2 portrays the sexual dimorphism in height.

As is the case with other dimorphisms in this class, an individual's height is affected by genes, nutrition (Perkins et al., 2016), disease (Stephensen, 1999) and a number of other less well-studied factors such as fluid intake and pregnancies prior to growth maturity (approximately age twenty); indeed, there are more than 500 variables involved in determining height (Wood et al., 2014).

What is true of height applies *mutatis mutandis* to a large number of differences between women and men. For example, women, *on average*, have slightly smaller brains, but brain size also varies on a spectrum. As with height, this difference connects with nothing intrinsically important. Albert Einstein, for example, had a slightly smaller than average size brain (1,230 grams; the average adult brain weighs about 1,400 grams). Here, a finer-grained analysis is needed such as the complexity of cellular and molecular organisation of neural connections (synapses), that is, what is going on in the brain during computational activity. Also, frontal lobe and grey matter volume, where dense neural cell bodies and synapses are found, has an impact on mental acuity and processing. *Superficial, gross differences* such as brain size seldom provide useful information.

From the numerous differences of this kind, I offer a few examples of the more often discussed. Females, *on average*, score lower on some IQ tests but here also there is a significant overlap of the curves, meaning many women

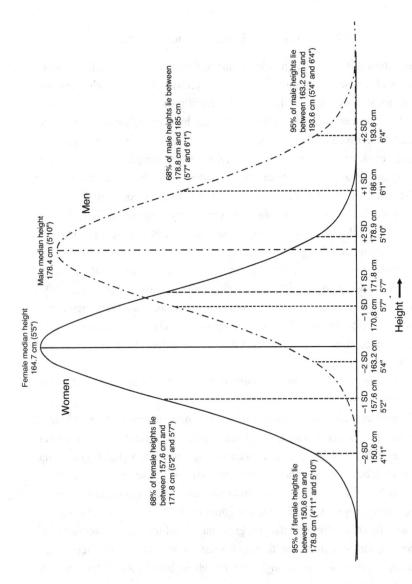

Figure 2 Sexual dimorphism in height

Note: The highest point on the curve is the average (mean) of the populations sampled. The spread is expressed as a standard deviation (SD). Men are *on average* taller than women. As the overlapping area shows, about one-third of the tallest women are taller than about one-third of the shortest men.

score higher than many men. Moreover, the *average* difference is at most four points, which, for any practical purpose, is a trivial difference. Strength and endurance reflect differences in testosterone and oestrogen levels and cycles; they are Gaussian distributed. The research on oestrogen and strength and risk of muscle injury is remarkably mixed (see Cauley, 2015; Deie et al., 2002; Frontera et al., 1991; Häkkinen and Pakarinen, 1993; Hansen, 2018; Kitajima and Ono, 2016). Women have advantages and disadvantages *on average*. Oestrogen's effects on endurance, for example, are an advantage (Bam et al., 1997; Lepers 2019). John Archer (2019) provides a quantitative overview of sex differences in human psychological attributes and considers the evidence for their possible evolutionary origins.

The germane point is that these differences do not provide an ERS-based rationale for unequal treatment within a cooperative social framework. In employment, for example, if strength is the important criterion, many women will outperform a significant proportion of men. Ultimately, equality requires that decisions be made based on requirements, not on sex,[20] regardless of the evolutionary origins of the difference or their continuing importance to ERS in non-human animals. Humans who participate in social groups enhance individual ERS through cooperation; that, as set out in Section 4.4, requires maximal individual liberty and equality. That there were female hunter-gatherers in early America (Haas, 2020) underscores the point that relevant attributes, not sex, determine the collective enhancement of individual ERS in social groups.

The foregoing suggests that sex differences, where they exist, fail to justify social policies of unequal treatment and, specifically, do not justify privileging males. The qualifier, 'where they exist', is important because sex-difference research is riddled with methodological, conceptual and logical flaws (Fine, 2010; Jordan-Young, 2010). This qualifier also applies to the discussions in the next section, even though the differences discussed are of a different kind.

6.3 Reproductive Sexual Dimorphisms

Reproduction-related sexual dimorphisms are different than those just discussed. Reproduction-related differences and ERS are inextricably intertwined. Focussing on humans, to state the obvious, a female has a vulva, vagina, clitoris, fallopian tubes, ovaries and a uterus; a male does not. A male has a penis, testes, a scrotal sac and a prostate gland; a female does not. A female ovulates, menstruates and can get pregnant and lactate; a male does not. A male produces sperm, prostatic fluid and seminal fluid; a female does not. These are differences in reproductive 'equipment' and 'products'; there are also important

---

[20] Gender has become culturally laden. I use the term 'sex' as a biological designation.

reproductively oriented differences in hormones, cognition and behaviour. All these are directed towards enhancing ERS.

Although males *qua* males and females *qua* females have similar reproductive interests, achieving those interests results in mating conflict among and between males and females. Trivers' (1985) rule highlights a clear aspect of these conflicting interests:

> The single most important difference between the sexes is the difference in their investment in offspring. The general rule is this: females do all of the investing; males do none of it. (p. 207)

The rule is indeed 'general' and, as one would expect, exceptions are numerous. The rule applies to humans but not in the extreme manner expressed; males do invest in offspring but rarely to the extent of females.

Unqualified, the ERS of males is enhanced by copulation with many females. Achieving this often involves male aggression and coercion (Buss, 2016, 2017; Clutton-Brock and Parker, 1995; Gavrilets et al., 2001; Smuts, 1992; Smuts and Smuts, 1993; Stumpf and Boesch, 2010). The male's cost of this behaviour is low. The cost of reproduction for females is much greater. Pregnancy involves physiological costs, and rearing an offspring to a reasonable probability of independent survival requires time, resources and risks. Hence, for a female, being choosy about mates and the timing of pregnancy are important elements in enhancing ERS. In general, consequently, persistence is an advantage for males and resistance an advantage for females. Success is enhanced by possessing characteristics that females find attractive – often signalling genetic superiority. Coercion through perseverance or outright aggression is beneficial to males. This results in selective pressure on males to evolve behavioural characteristics signalling genetic superiority and characteristics of persistence, and on females to evolve characteristics of resistance and mate selectivity (Rowe et al., 2005). This exposition only scratches the surface of the complexity of mating behaviours and the evolutionary dynamics involved but it is sufficient to convey the conflicts involved in reproductive behaviour (for an evolutionary-hormonal analysis of strategic modulation of reproductive effort by females, see Vitzthum, 2008).

Recently, it has also become clear that the foetus and infant are active agents in reproductive strategies. In the womb, the interaction of mother and foetus is significant. There is a conflict over resources. There is a selection dynamic related to allelic survival at certain loci – imprinted loci (Abramowitz and Bartolomei, 2012; Blunk, 2019). The mother has two alleles at each loci; the foetus has only one of those two – the other allele being paternal. Hence, the mother and foetus differ. There can be a complicated epigenetic-level conflict regarding allelic expression or even survival into the next generation (Haig, 1993, 2003, 2019; Mochizuki et al., 1996; Mora-Garcia and Goodrich, 2000; Spencer et al., 1998; Zeh and Zeh, 2000, 2002).

According to Hurst (1997), at least thirteen other theories, ranging from defence against ovarian tumours to chromosome surveillance, have been proposed to explain the evolution of this genomic imprinting. None of these alternative hypotheses seems sufficiently general to explain the diversity of patterns associated with parent-of-origin gene expression. By contrast, both whole genome and locus-by-locus analyses of mammalian imprinted gene expression have revealed a pattern that is highly concordant with intragenomic conflict playing a major role in maternal–foetal interactions. Most imprinted loci exert strong effects on foetal growth and development through expression in the trophoblast and placenta. In the majority of cases, alleles at paternally expressed loci act to promote maternal resource transfer to the foetus, whereas alleles at maternally expressed loci downregulate paternal genome activity. The conflict hypothesis also predicts that imprinting should be restricted to live-bearing species because only with viviparity (development *within* the female body) does the embryonic paternal genome have the opportunity to directly manipulate the mother's reproductive physiology. In support of this prediction, imprinting in plants is limited to angiosperms, whose reproductive mode is analogous to animal viviparity, with embryos being nourished by the endosperm for an extensive period after fertilisation (Mora-Garcia and Goodrich, 2000). Moreover, parent-of-origin gene expression does not occur in drosophila, birds (O'Neil et al., 2000) or the platypus (Killian et al., 2000), all of which lay eggs before or soon after fertilisation (development *outside* of the body).

Moreover, reproductive conflict does not end with birth. Once born, the infant is in a struggle for survival, a struggle where success is entirely dependent on others, especially the mother. Evolution is a dispassionate dynamical process, leading to the preservation of characteristics that favour ERS and the elimination of those that frustrate it. If, for instance, a propensity to nurture a second infant while still nurturing the first results in a high probability of losing both, that propensity will diminish in the population. Mothers evolved to make harsh assessments, among which is assessing the costs/benefits of putting time, emotions and resources into nurturing an infant. Making the wrong choice lowers one's ERS (Trivers, 1972, 1974). The infant's ERS, therefore, depends on making itself 'a good bet', and infant-behavioural characteristics to that end have evolved. Those that are lacking will have a lower probability of survival; a colicky infant, for example, is at risk (Levitzky and Cooper, 2000). Numerous propensities favouring reproduction have evolved, as have emotions and dispositions favouring nurturing and protecting offspring, but underlying these is a harsh ERS cost-benefit selection-based process. Infanticide and infant abandonment play a role in ERS. If in specific circumstances abandoning or killing an infant enhances the ERS of the mother, propensities for such behaviours can be expected to evolve (Hrdy, 1979, 2000, especially chaps. 12 and 14; Trivers, 1974).

The reproductive conflict among and between males, females, infants, mothers and foetuses and the corresponding evolved reproductive strategic propensities are clear; pretending that the biology is not what it is will result in failed social fabrics. Male domination, control and coercion of females are widespread among primates, with some notable exceptions such as bonobos (Muller and Wrangham, 2009; Smuts and Smuts, 1993; Stumpf and Boesch, 2010; Swedell et al., 2014). The challenge for males is attracting or coercing a female to mate and then protecting his reproductive investment (protection of the pregnant female and offspring from predators, resource insufficiency and other males). Decisions about protection require certainty about paternity. Patriarchy is a means of achieving a high degree of paternity-confidence as well as controlling mating and protecting a reproductive investment. Barbara Smuts (1995) has provided a comprehensive and compelling analysis of the evolution of patriarchy in humans (see also Hrdy, 1997). Smuts (1995) postulates six factors, presented as hypotheses in rough temporal order that 'influenced the evolution of human gender inequality [patriarchy]' (p. 20).

In many respects, her analysis, as well as Hrdy's, is a tale of males appropriating the emerging cooperative social structure, an arrogation that was accelerated by the emergence of language (allowing the formation of ideologies) and the rise of agriculture. It is an example of evolved propensities, which Smut describes in detail in the non-human primate world, that enhance ERS in one environment but diminish it in a changed environment, such as social cooperation, which is itself a product of evolved propensities. Consequently, the central issue is whether, in a social, cultural and, hence, moral context, the ERS of members is enhanced by deliberately overriding propensities that have resulted in patriarchy. On contractevolist theory, what needs to be demonstrated is that sexual equality enhances ERS in cooperative societies when compared with patriarchal societies.

The essence of patriarchy is conflict and control, principally of female reproductive behaviour by males, although the ramifications are much broader. As documented in this section, a wealth of knowledge about the reproductive behaviours of non-human primates and a host of animals is available, as are robust evolutionary explanations of those behaviours. Males secure reproductive opportunities through persistence, coercion and control directed at females.

Females respond with strategies of selective resistance and fertility control. These are evolved propensities and sexual selection has resulted in a metaphorical 'arms race' in which males evolve new strategies or refine existing ones for enhancing mating, and females evolve or refine strategies of resistance in response (Clutton-Brock and Parker, 1995; Rowe et al., 2005). This dynamic suggests that the evolved propensities of females are as important as those of males to the ERS of both sexes.

In humans, sexual behavioural propensities evolved in an environment where paternity, mating successes and investment protection mattered a great deal to male ERS. Human ERS is considerably more complicated than the number of inseminations. This is true throughout the animal kingdom as George C. William's (1966) noted: '[T]he maximization of individual reproductive success will seldom be achieved by unbridled fecundity' (p. 161). It is amplified in the human case because humans have altricial offspring, after a lengthy and dangerous gestation. Consequently, male ERS requires greater investment in infant rearing. This greater investment makes assurance of paternity important, which requires controlling access through successfully warding off male competitors and restricting female mating. Patriarchal social fabrics achieve both.

The advantages of cooperation among males to enhance their reproductive success were obvious, especially with existing propensities for persistence, coercion and control directed at females. Collusion among males reduced the need for male–male mating conflict, increased compliance of females and improved the probability of paternity. Barbara Smut's (1995) six hypotheses about the elements of the evolutionary origins of patriarchy align well with this narrative, as well as explaining why similar coalitions among females did not emerge. All the elements of patriarchy are in place. There is a division of interests and propensities by sex and the emergence of male cooperation to amplify them to male advantage. Smut's sixth hypothesis provides yet another vehicle for patriarchy: 'The evolution for the capacity for language allowed males to consolidate and increase their control over females because it enabled the creation and propagation of ideologies of male dominance/female subordination and male/female inferiority' (p. 19).

Regrettably, this set humans on a path to suboptimal fitness in terms of ERS. Sewall Wright, in a number of publications beginning in 1931, introduced the concept of adaptive landscapes (Wright 1931, 1932). An adaptive landscape is a mountainous topography with the height of the mountains mapping onto average fitness (ERS in this case). The higher the peak, the greater the population fitness. A population beginning in a valley will climb a fitness mountain to its peak – its optimal fitness in that environment. On which mountain peak a population finds itself depends on where in the valley it started. In this model, a population could climb to a fitness peak that is lower than others in the landscape. Thus, it is suboptimal; had it climbed a different mountain, it would have achieved greater fitness. Once at a peak, however, moving to a different mountain requires descending the one already climbed. That is, fitness must decrease in order to achieve a more optimal peak, and this will not occur. Wright (1932, 1977) developed a 'shifting balance theory' that would allow populations to cross valleys to other peaks through genetic drift, gene interactions and subdivisions of populations. Shifting balance theory is controversial (Coyne et al., 1997; Coyne et al., 2000; Hodge, 2011; Peck et al., 1998; Wade and Goodnight, 1998).

Although many underlying propensities have a long evolutionary history, contemporary patriarchy is a function of the evolution of societies. Two important transformations in this evolution were the evolution of language and the advent of agriculture. Although genetical evolution has continued, cognition and cultural evolution have increased in importance over the last 50,000 years. The fabric of societies, even small groups, is more a culture-driven process than genetic or epigenetic, the latter two having laid down most of the determinants. Hence, the climb up the fitness mountain to patriarchy is mostly culture-driven, with the biological propensities partially determining the available mountains to climb. Ron Vannelli (2015) provides a useful, sustained treatment of the interconnection of evolution and social factors.

Patriarchy exploits the human propensity for cooperation but only partially, and it is this that makes it suboptimal in enhancing the ERS of the members of the society. Males are the major beneficiaries of enhanced ERS. There are benefits for females (ideally) – protection from aggression towards her and her offspring and greater resource security, for example – but females pay a significant price for these benefits (reduction in mate variation, for example). Moreover, far too frequently the reality falls short of the ideal and females and their offspring are abused rather than protected. These are impediments to the ERS of both males and females.

Compared to a social fabric of equality of all members – one where all members are partners in the cooperative venture to enhance ERS – patriarchy fares poorly. It is suboptimal in achieving enhanced ERS for its members, even those it is supposed to advantage: males. As a turn of the utilitarian maxim, the greatest enhancement of ERS for the greatest number results from the cooperation of all, not from the cooperation of one sex in securing the bondage of the other. For example, a society in which females participate in determining the number, spacing and resource requirements of offspring, as well as methods that maximise knowledge of paternity, yields considerable enhancements of male ERS. Similarly, patriarchs might postulate that males are better equipped to secure resources, although hunter-gatherer societies suggest that this is very environment-dependent; but in the current environment, many women are better equipped physically, intellectually and emotionally than their current male partners to secure the needed resources. Hence, greater equality of opportunity to secure resources along with greater equality to fill specific roles in the reproductive process increases the ERS of males and females. In short, it increases the ERS of males compared with their ERS in paternalistic societies. Cooperative strategies to resolve reproductive conflicts enhance ERS. Smuts' second hypothesis highlights this for male–male reproductive conflicts:

> Over the course of human evolution, male-male alliances became increasingly well-developed. These alliances were often directed against females, and they increased male power over females. (Smuts, 1995, p. 13)

This is the essence of patriarchy, but it limits the fruits of reproductive success; both males and females can reap greater ERS by cooperation across the divide of sex-based reproductive conflicts. Equality is the route up the adaptive mountain to a greater fitness peak for males and for females.

This equality, however, can only be crafted and maintained by careful attention to inherent biological propensities. Culture, through a social fabric, can shape these but not remove them. We are animals; we are sexual beings. No amount of wishful thinking can alter that reality. The more we understand our animality, the greater will be our success in harnessing it.

7 Individual Sovereignty and Copulatory Choices

Contractevolism is a biologically based, naturalistic moral theory that espouses the primacy of ERS within a social contract. As such, there is a danger that it might be construed as entailing that individuals have an obligation to maximise their ERS, with obvious implications for moral values regarding copulatory behaviour. I have been careful to cast contractevolism in terms of maximising *opportunities* for ERS. An obligation on a society to ensure that its codes maximise opportunities for individual ERS respects individual liberty (sovereignty) to choose whether to avail oneself of the opportunities. Restrictions on this liberty must be based solely on achieving the goal of maximising the opportunities for ERS for all members, one element of which will be the maximum stability of the social structure.

Maximum opportunities for individual ERS in a non-social environment will be achieved by maximum sovereignty over one's copulatory behaviour. Some aspects of that sovereignty will need to be surrendered by members of a cooperative, stable society. Given the obvious link between maximum sovereignty over one's copulatory behaviour and maximum opportunities for ERS, justifying social/moral rules restraining or requiring specific copulatory behaviours faces a high hurdle.

Two interrelated things require attention. First, many copulatory desires and behaviours flow from evolutionary propensities and have deep hormonal or emotional drivers. Second, changing environments require changing social/ moral rules. This applies to non-sexual propensities as well, such as xenophobia. The propensity to fear strangers persists today (Aarøe et al., 2017; Öhman and Mineka, 2001), but it needs to be understood and socially mitigated (see McEvoy, 2002) because, in the modern world, it works against social stability and inclusivity. The focus of this section, however, is on copulatory behaviours because of their intrinsic connection to ERS.

7.1 Evolutionary Cost-Benefit Dynamics

The contractevolist analysis begins with a recognition that humans are flooded with sexual hormones and sexual emotions, which promote *and regulate* copulation – regulate because, even in the absence of any social considerations, there are evolutionary costs and benefits of all reproductive behaviours.

One example of the complex cost-benefit dynamics is the evolution of cooperative breeding (alloparenting). Helping close kin raise their children can be explained in many cases in terms of inclusive fitness, but helping raise the offspring of non-relatives seems prima facie contrary to one's reproductive interests. Nonetheless, it occurs broadly in nature (see Clutton-Brock, 2002; Clutton-Brock et al., 1999; Clutton-Brock et al., 2001; Martin et al., 2020; Perry and Daly, 2017; Riedman, 1982; Rood, 1990; Russell, 2004). The best evolutionary explanation is based on the recognition that, in cooperative breeding species, everyone benefits from the cohesive behaviour of the group. By acting as a group to improve the resources (e.g., food and shelter), defence from predation, respite from wearying activities and so on, *everyone* benefits in terms of a longer, healthier life, which in turn directly benefits helpers because it increases their chances of successful breeding and nurturing at some future time – a rising tide floats all boats.

Monogamy (also pair bonding and stable breeding bonds) is another example where the evolutionary cost-benefit dynamic will *under certain conditions* constrain sexual behaviour (see Chapais, 2008; Langergraber et al., 2013; Lukas and Clutton-Brock, 2013). In humans, a weak variety of monogamy was driven by the need for extended paternal investment, which 'shifted competition between males for mates, which was potentially destructive for the group, to a new dimension which is beneficial for the group – competition to be a better provider to get better mates' (Gavrilets, 2012a, p. 9927). An additional factor is 'reduced female density and limited potential for males to guard more than one female' (Lukas and Clutton-Brock, 2013, p. 529; see also Langergraber et al., 2013). These are robust hypotheses demonstrating that constrained copulatory behaviour can be explained biologically.

An important example of the cost-benefit element of evolutionary dynamics often results in individuals not having as many offspring as possible. Many species have evolved mechanisms to reduce reproduction when resources are scarce or territory becomes crowded. Richard Alexander (1987, p. 179) has, with clarity and deftness, dispatched the idea that human reproductive success is measured by the number of offspring. Moreover, Trivers (1972) has demonstrated theoretically that an individual's ERS can be achieved by mating or parenting. Experimental evidence suggests that this is not an either-or strategic situation (see Heath and Hadley, 1998). The message of these examples is clear: evolutionary dynamics, even in asocial contexts, has resulted in significant constraints on copulatory behaviours.

Biology also shapes our copulatory behaviour in subconscious ways. Hormonal sexual 'signals' have evolved which shape our desires and actions; many, perhaps most of the important ones, are produced and detected unconsciously, such as ovulation. Miller et al. (2007) found that lap dancers received significantly larger tips when they were ovulating, suggesting that some communication of ovulation was occurring but not consciously (see also Miller and Maner, 2010); and Law-Smith et al. (2006) found that facial features were different when high urinary oestrogen levels were present – men reported finding the faces more attractive. Research on genes in the major histocompatibility complex (MHC, known in humans as human leukocyte antigen loci) suggest that odours, and possibly facial features, signal aspects of the MHC and allow detection of those with similar or different MHC, which can allow kin identification and immunocompetence assessment of mating combinations and promote overall genetic diversity (Havlicek and Roberts, 2009). These and many other signals regarding genetic and hormonal characteristics, as well as fertility cycles, play an important role in mate selection and timing.

This research is important not because it demonstrates that biology *determines* mate choice and sexual behaviour – it doesn't – but because failure to take it into account in personal and social decision-making increases the probability of poor and unjustifiable decisions. Human mate choice is complex, involving many variables: some social, some contextual and some biological, with emotions generated by each. Research on these provides some guidance on effective decision-making, most especially in social policy and moral norms; individual choices will vary widely and not always be comprehensible – perhaps even to the decision-maker. By way of analogy, in many areas of medicine, where lifestyle, social expectations, biology and other factors are involved in illness, failure to take into account what we know biologically can lead to misdiagnosis, ineffective therapy and even blaming the victim, with resultant psychological effects.

Finally, although understanding copulation-related hormones, emotions and behaviours as evolved phenomenon, and the constantly changing evolutionary cost-benefit ratio of copulatory behaviours is the starting point, determining what, if any, social contract benefits and costs there are in promoting, modifying or prohibiting the behaviours draws also on psychological and sociological knowledge and theories as well as current circumstances. With respect to the latter, for example, the availability of effective methods of contraception control potentially changes the benefit-to-cost ratio of many copulatory behaviours.

7.2 Social Intervention

Contractevolism requires that any intervention – prescriptive or proscriptive – rests on establishing that the reciprocal/cooperative underpinnings of the society require it, and the costs of diminished individual liberty are outweighed by those social benefits and the resultant enhancement of opportunities for maximising ERS.

Prescribing and proscribing are coercive. Although coercion of an individual or group of individuals by another individual or group is seldom a justifiable reduction of liberty, social coercion is frequently justifiable. Children are coerced to attend school and their parents coerced to ensure that they do so. In times of demonstrable need, citizens are drafted into the service of defending their society. Citizens can be coerced into quarantine during an epidemic. Although there are rough edges to such coercions and contests over evidence, and justifications abound in specific cases, the necessity of specific cases of social coercion remains. The limits and justification of coercion in the case of copulatory behaviours are illustrative of the broader contractevolist limits and justifications of social coercion.

Social proscription of copulatory behaviour is more common than prescription. Most instances of prescription today would seem akin to Margaret Atwood's dystopian society in *The Handmaid's Tale*. Although physically forced sexual intimacy within marriage has been socially condoned, its acceptance has been mostly exorcised from secular European and English-speaking legal frameworks. Regrettably, it persists in the doctrines of some religions and creeds, and in societies based on them.

Prohibition, by contrast, of some copulatory behaviours is uncontroversial: copulation with minors and the cognitively challenged and rape, for example. In these cases, meaningful consent is absent or dubious, pointing to the absence of consent as being the key factor in social proscription. Prostitution, on the other hand, is a more interesting case of the proscription of a copulatory behaviour because meaningful consent to participate or refrain can, in principle, be given. Prostitution, *simpliciter*, is employment.

The uncoerced purveying of one's physical or cognitive attributes for payment (employment) is widely accepted. There are always costs and benefits involved. Sports can ravage one's physiology. Aspects of police and military work, firefighting and the like are dangerous, often putting life and limb on the line. Miners face many dangers, some long-term. The psychological stress of some employment is all too often destructive – on family and social life, for example. Employment can trap, frustrate and discourage some people. The response to such costs has not been to prohibit vast swaths of the employment landscape but to attempt to eliminate or at least mitigate the harms. The obvious benefit is income, but there are also idiosyncratic benefits: personal gratification, for example.

Attitudes to, and laws proscribing, prostitution have varied.

> [A]fter all these centuries [from before Babylonian/Old Testament times] . . .
> there has been no appreciable alteration in the reaction of society as a whole
> to prostitution. Publicly the prostitute was denounced, just as she is to-day;
> privately she was supported and encouraged. (Scott, 2013, pp. 63–4)

Religious views have ranged from coercion,

> [A]ccording to Herodotus [book 1, chap. 199], the women of Babylonia were
> required to sit in the temple of Mylitta until some men claimed the right to
> have intercourse with them. In other words, each woman was required to
> become a temporary prostitute, the fee paid by the man constituting an
> offering to the goddess presiding over the temple. (Scott, 2013, p. 55)

to condemnation, derived in Roman Catholicism from Augustine's concept of original sin:

> Erotic desires and passions were not part of God's original plan for our sexual
> lives, as the Pelagian heretic Julian of Eclanum taught, but a consequence of
> sin. According to Augustine, sin caused a disjunction between our bodies and
> wills, mirroring the split between God's will and our wills – our bodies no
> longer obey reason and the will but are moved by lust. (Wood, 2000, p. 36)

Elaine Pagels (1989) provides an excellent critical appraisal of the views of Irenaeus and Augustine and on the creation of the concept of 'original sin' and its connection to sexual desire.

Feminist thought is divided. Some argue that it is yet another example of male subjugation of women (see Barry, 1995; Dworkin, 1981, 1997; Jeffreys, 1997; MacKinnon, 1987, 1989). Ronald Weitzer (2012) accurately sums up this literature: 'Radical feminism sees prostitution as the quintessential form of male domination over women – the epitome of women's subordination, degradation, and victimization' (p. 211). Many other feminists favour prostitution's legalisation with effective regulation on the grounds that it is prohibition that stigmatises and victimises women (Frances and Gray, 2007) and, importantly, prohibition compromises the health and safety of women involved.

Trafficking, exploitation and involvement of those below the age of consent are clear examples of non-consent, and social proscription is justified to promote liberty, not undermine it. This is not unique to prostitution; where there is commerce, there will be scoundrels bringing misery and leaving tragedy in their wake. The history of prostitution, however, suggests that the harms are not intrinsic, as does widespread casual copulation, where no money changes hands but the behaviour is the same. Moreover, the mitigation efforts in a number of countries have lighted the way to successful mitigation of harms (Weitzer, 2012).

There are risks associated with prostitution – sexually transmitted diseases, unwanted pregnancy and increased risk of certain cancers (Gabovac, 2020), for example – but these also apply to casual copulatory behaviour and serial monogamous behaviour. A simple measure – using a condom – significantly mitigates many risks. Moreover, risk-taking is not unique to prostitution. Many people choose risky activities, from wilderness heli-skiing to skydiving; even driving a car has significant risks, and those who choose a profession in frontline emergency services and the military accept those risks for remuneration. Life involves risks, some of which we get paid for taking.

Proscribing or denigrating prostitution where copulation is consensual minimises liberty. Contractevolism cuts to the chase in this respect. Arguments about the sanctity of the human body along with notions of self-respect, dignity and self-worth are emotive – perhaps expressing one's own choices and one's personal identity; within contractevolist theory, they provide an insufficient justification for proscription or denigration. Copulation is essential to maximising opportunities for ERS. The social contract is obliged to maximise these opportunities, not to restrict the freedom to take such opportunities. It is also obliged to ensure that choices are made freely. *Contractevolism entails a strong principle of liberty, such that restriction of any uncoerced sexual behaviour requires a high standard of justification.*

For contractevolists, sexual desire and copulation are an essential component of ERS, without which the species would become extinct. That is why both are important elements of contractevolist moral theory. Prostitution is not about commodification of the body – such commodification abounds in many realms, especially commercial realms – but about commodification of carnal copulatory desire: animality incarnate. There will be many idiosyncratic and prudential reasons for individuals to temper this desire; we all do and, to varying degrees and for changing contextual reasons, have for tens of millennia. Contractevolism, however, embraces the moral goodness of copulatory desire and action while condemning those things that restrain individual liberty – for example, exploitation and coercion of those involved. That alone is a daunting social challenge.

8 Contractevolism and Overpopulation

Over the last 12,000 years, humans have increased the opportunities for ERS. The domestication of plants and animals about 11,000–12,000 years ago (Schlegel, 2018; Vigne, 2011) began this process. In the last 500 years, many other factors have contributed to the growth in population, most are science-based: industrialisation, advances in health care (disease and infection control, mitigation of the perils of childbirth, reduced childhood mortality and better nutrition, for example), advances in agriculture (fertilisation, irrigation, weed and pest control) and other changes such as urbanisation and increased and diversified energy resources. These advances have been bumpy but have steadily increased the opportunities for ERS and with it the size and distribution of human populations. Superficially, this is morally praiseworthy on contractevolist theory. It has, however, led to overpopulation, by which I mean a population size that places unsustainable pressure on the environment, a consequence that appears to be the *reductio ad absurdum* of contractevolism.

Things are more complicated, however, than this sketchy analysis suggests. An initial response, also sketchy, is that behaviours and decisions that lead to population collapse do not maximise opportunities for ERS even though they may seem to do so in some temporal slice of evolution. Although full development of this response is not possible here, the core elements are set out.

Evolutionary dynamics, in principle, result in balance. Predators increase and deplete their prey; predators starve, their population declines and the prey rebound; an oscillating balance ensues. Selection is constantly acting through these cycles. In some species, fertility declines as population density increases. In this way, maximal ERS is achieved. Humans, however, have dramatically decreased predation, either by carnage, by changing ecosystems (e.g., clearing woodlands, draining swamps) or by thwarting pathogens. Moreover, humans have successfully increased agricultural yields, most recently through genetic modification. In short, humans have disrupted the dynamics that operated since the emergence of life.

Disruption, however, is not the problem; it happens all the time and 'nature' oscillates towards a new balance. It is the magnitude, relentlessness and temporal rapidity of the disruption. The increase in opportunities for ERS that these disruptions have created seems, on contractevolism, morally good. To the extent, however, that increased opportunities for ERS during a brief window of time imperil longer-term opportunities for ERS, they have to be considered morally wrong (evil). The well-known way through the horns of this moral dilemma is to recognise that some actions, viewed in the short term (a few generations at a time), appear to be morally good but, cumulatively and in the longer term, are not.

That appears to be a promising solution to the moral dilemma, but is it faithful to the evolutionary stance on which contractevolism is built? A pivotal evolutionary observation is that all species become extinct; only arrogance blinds us to the inevitability of human extinction. Van Valen's (1973) law gives dynamical substance to the claim by positing that evolution is a zero-sum game. In order for one species to win – increase its ERS – one or more others must lose. This can be best understood using only two species, but it is generalisable. Van Valen argued that one species α can only increase its ERS at the expense of the ERS of a coexisting species β. However, evolution also will drive β to increase *its* ERS and that will be at the expense of α. At any given time, the 'total fitness' (fitness of $(\alpha + \beta)$) remains the same. A predator–prey relationship illustrates this. The ERS of a predator species will be increased if a new advantageous characteristic emerges. But, over time, the prey species will evolve a defence. Again, over time, the predator will find a new 'trick'. And on it goes. Van Valen called this the 'Red Queen's Hypothesis' after an incident in Lewis Carroll's *Through the Looking Glass*. Alice and the Red Queen were running but staying in the same place – 'Now, *here*, you see, it takes all the running you can do, to keep in the same place', the Red Queen remarked (Carroll, 1871/2001, p. 179). Others have likened it to an arms race.

Each species is on a treadmill – changing to adapt to the surrounding changes. The treadmill ends for a species with the extinction of one or both of the species, and that happens when the effective environment diminishes below a minimum level of species support. If this is a dynamic of evolution, then one should bite the bullet and accept the treadmill, make it work to human advantage for as long as possible and accept that extinction lies at the end; it is inevitable. There is no moral dilemma; ERS – short term and long term – justifies staying our current course.

There is something unsettling and unsatisfying about this simple appeal to evolutionary dynamics. In its simplicity, it ignores the magnitude, relentlessness and temporal rapidity of human disruption. No longer do humans advance their ERS and other species in response advance theirs. Humans are doing almost all the advancing – the microbial world possibly excepted; it's like having a finger on the evolutionary balance scale.

The result is that we are temporally advancing our extinction – hence, temporally shortening our ERS. Moreover, we are taking a vast number of species with us on the rapid path to extinction. There is no metaphorical arms race; humans are capturing all the 'total fitness' available and, thereby, crossing a threshold beyond which 'total fitness' begins declining. At, or close to zero, most species, including humans, become extinct.

It is not only the very long-term ERS, however, that is at risk but the present and very near future ERS. There are multiple indices: resource depletion, climate change with related health diminution, valuable (even from a human-centric perspective) species depletion, changing patterns of rainfall and temperature, and the list goes on. If we continue on our current course, we may dominate the earth, but it will be domination of a landscape akin to the American and Canadian prairies in the 1930s, after the dust-bowl debacle; then the collapse of the human species.

Focussing on overpopulation as a major element, there is a tipping point beyond which population size is unsustainable. It is, of course, a mug's game to try to determine the tipping point and it is almost certainly a moving target. The Malthusian tipping point, so important to Darwin's theorising, demonstrates this; science and technology have increased resources faster than populations have grown. Environmental degradation has become a pivotal threshold; pollution, in all its forms, and non-food resource depletion take us close to the threshold. Humans are ingenious and will no doubt be able to put off the day of reckoning, but contractevolism requires that the impact of overpopulation and environmental degradation be included in assessments of a social fabric's success in maximising opportunities for ERS.

Also important, population size matters to social functioning: productive sociality and cooperation. The sense of citizenship, obligation, cohesion and cooperation is stressed. Conservation biologists have, for obvious reasons, focussed, through population viability analysis (Beissinger and McCullough, 2002), on the lower boundary of viability – that is, the point below which population size endangers viability. The upper boundary has received less attention (Dunbar et al., 2018; Zhou et al., 2005).

References

Aarøe, Lene, Michael Bang Petersen and Kevin Arceneaux (2017) 'The Behavioral Immune System Shapes Political Intuitions: Why and How Individual Differences in Disgust Sensitivity Underlie Opposition to Immigration', *American Political Science Review* 111 (2): 277–94.

Abramowitz, Lara K. and Marisa S. Bartolomei (2012) 'Genomic Imprinting: Recognition and Marking of Imprinted loci', *Current Opinion in Genetic Development* 22 (2): 72–8.

Alexander, Richard D. (1974) 'The Evolution of Social Behavior', *Annual Review of Ecology and Systematics* 5: 325–83.

(1987) *The Biology of Moral Systems*. New York: Aldine de Gruyter.

Andrade, M. C. B. (1996) 'Sexual Selection for Male Sacrifice in the Australian Redback Spider', *Science* 271 (5245): 70–2.

(2003) 'Risky Mate Search and Male Self-Sacrifice in Redback Spiders', *Behavioural Ecology* 14 (4): 531–8.

Archer, John (2019) 'The Reality and Evolutionary Significance of Human Psychological Sex Differences', *Biological Review* 94 (4): 1381–415.

Axlerod, Robert M. (1984) *The Evolution of Cooperation*. New York: Basic Books.

Axlerod, Robert M. and William. D. Hamilton (1981) 'The Evolution of Cooperation', *Science* 211 (4489): 1390–6.

Baldwin, Thomas (1992) *G. E. Moore*. London: Routledge.

Bam, J., T. D. Noakes and S. C. Dennis (1997) 'Could Women Outrun Men in Ultramarathon Races?', *Medical Science of Sports Exercise* 29 (2): 244–7.

Barry, Kathleen (1995) *The Prostitution of Sexuality*. New York: New York University Press.

Beck, John H. (2007) 'The Pelagian Controversy: An Economic Analysis', *The American Journal of Economics and Sociology* 66 (4): 681–96.

Beissinger, Steven R. and Dale R. McCullough (eds.) (2002) *Population Viability Analysis*. Chicago, IL: University of Chicago Press.

Bekoff, Marc (1995) 'Cognitive Ethology: The Comparative Study of Animal Minds', in William Bechtel and George Graham (eds.), *Blackwell Companion to Cognitive Science*, pp. 371–9. Oxford: Blackwell Publishers.

Bergstrom, Carl T. (2003) 'The Algebra of Assortative Encounters and the Evolution of Co-operation', *International Game Theory Review* 5: 2110228.

Binmore, Ken (1994) *Game Theory and the Social Contract, Volume 1: Playing Fair*. Cambridge, MA: MIT Press.

Birch, Jonathan and Samir Okasha (2015) 'Kin Selection and Its Critics', *Bioscience* 65: 22–32.

Bissonnette, Annie, Susan Perry, Louise Barrett et al. (2015) 'Coalitions in Theory and Reality: A Review of Pertinent Variables and Processes', *Behaviour* 152 (1): 1–56.

Blunk, Inga, Manfred Mayer, Henning Hamann and Norbert Reinsch (2019) 'Scanning the Genomes of Parents for Imprinted Loci Acting in Their Un-genotyped Progeny', *Science Reports* 9 (1): 654–71.

Bolton, Gary E. and Axel Ockenfels (2000) 'ERC: A Theory of Equity, Reciprocity, and Competition', *The American Economic Review* 90 (1): 166–93.

Buchanan, James (1975) *The Limits of Liberty: Between Anarchy and Leviathan*. Chicago, IL: Chicago University Press.

Buchanan, James and Gordon Tullock (1962) *The Calculus of Consent: Logical Foundations of Constitutional Democracy*. Ann Arbor: University of Michigan Press.

Buckle, Stephen and Dario Castiglione (1991) 'Hume's Critique of the Social Contract', *History of Political Thought* 12 (3): 457–80.

Buckner, Randy L. and Fenna M. Krienen (2013) 'The Evolution of Distributed Association Networks in the Human Brain', *Trends in Cognitive Science* 17 (12): 648–65.

Bulmer, Michael (1994) *Theoretical Evolutionary Ecology*. Sunderland, MA: Sinauer Associates Publishers.

Buss, David M. (2016) *The Evolution of Desire: Strategies of Human Mating* (4th ed.). New York: Basic Books.

(2017) 'Sexual Conflict in Human Mating', *Current Directions in Psychological Science* 26 (4): 307–13.

Calmettes, Guillaume and James N. Weiss (2017) 'The Emergence of Egalitarianism in a Model of Early Human Societies', *Heliyon* 3 (11): e00451.

Carroll, Lewis (1871/2001) *Alice in Wonderland & Through the Looking Glass*. Ware, UK: Wordsworth Editions.

Cauley, Jane A. (2015) 'Estrogen and Bone Health in Men and Women', *Steroids* 99 (Pt. A): 11–15.

Chapais, B. (2008) *Primeval Kinship: How Pair-bonding Gave Birth to Human Society*. Cambridge, MA: Harvard University Press.

Cheney, Dorothy L. (2011) 'Extent and Limits of Cooperation in Animals', *Proceeding of the National Academy of Sciences* 108 (S2): 10902–9.

Chidi-Ogbolu, Nkechinyere and Keith Baar (2019) 'Effect of Estrogen on Musculoskeletal Performance and Injury Risk', *Frontiers in Physiology* 9: 1834. https://doi.org/10.3389/fphys.2018.01834.

Clutton-Brock, Tim (2002) 'Breeding Together: Kin Selection and Mutualism in Cooperative Vertebrates', *Science* 296 (5565): 69–72.

Clutton-Brock, Tim H. and Geoff A. Parker (1995) 'Sexual Coercion in Animal Societies', *Animal Behavior* 49 (5): 1345–65.

Clutton-Brock, T. H., M. J. O'Riain, P. N. Brotherton et al. (1999) 'Selfish Sentinels in Cooperative Mammals', *Science* 284 (5420): 1640–4.

Clutton-Brock, T. H., A. F. Russell, L. L. Sharpe, G. M. McIlrath, S. White and E. Z. Cameron (2001) 'Effects of Helpers on Juvenile Development and Survival in Meerkats', *Science* 293 (5539): 2246–9.

Collier, John and Michael Stingl (2020) *Evolutionary Moral Realism*. New York: Routledge.

Coyne, Jerry A., Nicholas H. Barton and Michael Turelli (1997) 'Perspective: A Critique of Sewall Wright's Shifting Balance Theory of Evolution', *Evolution* 51 (3): 643–71.

(2000) 'Is Wright's Shifting Balance Process Important in Evolution?', *Evolution* 54 (1): 306–17.

Crisp, Roger (ed.) (2013) *The Oxford Handbook of the History of Ethics*. Oxford: Oxford University Press.

Darwin, Charles (1859) *On the Origin of Species*. London: John Murray.

Davies, Nicholas B., John R. Krebs and Stuart A. West (2012) *An Introduction to Behavioral Ecology* (4th ed.). Oxford: Wiley-Blackwell.

Dawes, Christopher T., James H. Fowler, Tim Johnson, Richard McElreath and Oleg Smirnov (2007) 'Egalitarian Motives in Humans', *Nature* 446 (7137): 794–6.

Deacon, Terrence W. (1997) *The Symbolic Species: The Co-evolution of Language and the Brain*. New York: W. W. Norton & Company.

Deie, Masataka, Yukie Sakamaki, Yoshio Sumen, Yukio Urabe and Yoshikazu Ikuta (2002) 'Anterior Knee Laxity in Young Women Varies with Their Menstrual Cycle', *International Orthopaedics* 26 (3): 154–6.

Dennett, Daniel C. (1995) *Darwin's Dangerous Idea: Evolution and the Meanings of Life*. New York: Simon & Schuster.

Dewey, John (1898) 'Evolution and Ethics.' Reprinted in M. H. Nitecki and D. V. Nitecki (eds.) (1993) *Evolutionary Ethics*, pp. 95–110. Albany: State University of New York.

Dugatkin, Lee Alan (1997) *Cooperation among Animals: An Evolutionary Perspective*. Oxford: Oxford University Press.

(1988) 'Do Guppies Play Tit for Tat during Predator Inspection Visits?', *Behavioral Ecology and Sociobiology* 23 (6): 395–9.

Dunbar, Robin I. M. and Richard Sosis (2018) 'Optimising Human Community Sizes', *Evolution and Human Behavior* 39 (1): 106–11.

Dworkin, Andrea (1981) *Pornography: Men Possessing Women*. New York: Putnam.

(1997) *Life and Death: Unapologetic Writings on the Continuing War Against Women*. London: Virago Press.

Egerton, Frank N. (2016) 'History of Ecological Sciences, Part 56: Ethology until 1973', *Bulletin of the Ecological Society of America* 97 (1): 31–88.

Eshel, Ilan and Luigi L. Cavalli-Sforza (1982) 'Assortment of Encounters and Evolution of Cooperativeness', *Proceedings of the National Academy of Sciences* 79 (4): 1331–5.

Fehr, Ernst and Klaus M. Schmidt (1999) 'A Theory of Fairness, Competition, and Cooperation', *Quarterly Journal of Economics* 114 (3): 817–68.

Ferguson, John (1956) *Pelagius: A Historical and Theological Study*. Cambridge: W. Heffer & Sons.

(1980) 'In Defence of Pelagius', *Theology* 83 (692): 114–19.

Fine, Cordelia (2010) *Delusions of Gender: How Our Minds, Society, and Neurosexism Create Difference*. New York: W. W. Norton & Company.

Frances, Raelene and Alicia Gray (2007) 'Unsatisfactory Discrimination, Unjust and Inviting Corruption: Feminists and Decrimilalisation of Street Prostitution in New South Wales', *Australian Feminist Studies* 22 (53): 307–24.

Frank, Steven A. (2000) 'Polymorphism of Attack and Defence', *Trends in Ecology and Evolution* 15 (4): 167–71.

Freud, Sigmund (1971) *The Complete Introductory Lectures on Psychoanalysis*, ed. and trans. James Strachey. London: George Allen & Unwin. (Originally published in German, 1916–17.)

Frontera, Walter R., Virginia A. Hughes, Karyn J. Lutz and William J. Evans (1991) 'A Cross-Sectional Study of Muscle Strength and Mass in 45- to 78-Yr-Old Men and Women', *Journal of Applied Physiology* 71 (2): 644–50.

Gabovac, Igor, Lee Smith, Lin Yang et al. (2020) 'The Relationship between Chronic Diseases and Number of Sexual Partners: An Exploratory Analysis', *British Medical Journal* 46 (2): 100–7.

Gauthier, David (1979) 'David Hume, Contractarian', *The Philosophical Review* 88 (1): 3–38.

(1986) *Morals by Agreement*. Oxford: Clarendon Press.

Gavrilets, Sergey (2012a) 'Human Origins and the Transition from Promiscuity to Pair-Bonding', *Proceedings of the National Academy of Sciences of the United States of America* 109 (25): 9923–8.

(2012b) 'On the Evolutionary Origins of the Egalitarian Syndrome', *Proceedings of the National Academy of Sciences of the United States of America* 109 (35): 14069–74.

Gavrilets, Sergey, Göran Arnqvist and Urban Friberg (2001) 'The Evolution of Female Mate Choice by Sexual Conflict', *Proceedings of the Royal Society of London B: Biological Sciences* 268 (1466): 531–9.

Grafen, Alan (1985) 'A Geometric View of Relatedness', *Oxford Surveys in Evolutionary Biology* 2: 28–90.

(1991) 'Modelling in Behavioural Ecology', in John R. Krebs and Nicholas B. Davies (eds.), *Behavioural Ecology*, pp. 5–31. Oxford: Blackwell.

Graur, Dan and Li Wen-Hsiung (2000) *Fundamentals of Molecular Evolution* (2nd ed.). Sunderland, MA: Sinauer Associates.

Haakonssen, Knud (1992) 'Natural Law', in Lawrence C. Becker and Charlotte B. Becker (eds.), *Encyclopedia of Ethics*, pp. 884–90. New York: Garland Press.

Haas, Randall (2020) 'Female Hunters of the Early Americas', *Science Advances* 6 (45): 1–10.

Haig, David (1993) 'Genetic Conflicts in Human Pregnancy', *Quarterly Review of Biology* 68 (4): 495–532.

(2003) 'Behavioural Genetics: Family Matters', *Nature* 421 (6922): 491–2.

(2019) 'Cooperation and Conflict in Human Pregnancy', *Current Biology* 29 (11): R455–8.

Häkkinen, K. and A. Pakarinen (1993) 'Muscle Strength and Serum Testosterone, Cortisol and SHBG Concentrations in Middle-Aged and Elderly Men and Women', *Acta Physiologica Scandinavia* 148 (2): 199–207.

Hamilton, W. D. (1964a) 'The Genetical Evolution of Social Behaviour I', *Journal of Theoretical Biology* 7 (1): 1–16.

(1964b) 'The Genetical Evolution of Social Behaviour II', *Journal of Theoretical Biology* 7 (1): 17–52.

Hammerstein, Peter (ed.) (2003) *Genetic and Cultural Evolution and Cooperation*. Cambridge, MA: MIT Press.

Hansen, Mette (2018) 'Female Hormones: Do They Influence Muscle and Tendon Protein Metabolism?' *Proceedings of the Nutrition Society* 77 (1): 32–41.

Harman Oren (2011) 'Birth of the First ESS: George Price, John Maynard Smith, and the Discovery of the Lost "Antlers" Paper', *Journal of*

Experimental Zoology Part B: Molecular and Developmental Evolution 316B (1): 1–9.

Hartl, Daniel L. and Andrew G. Clark (2007) *Principles of Population Genetics* (4th ed.). Sunderland, MA: Sinauer Associates.

Havlicek, Jan and S. Craig Roberts (2009) 'MHC-Correlated Mate Choice in Humans: A Review', *Phychoneuroendocrinology* 34 (4): 497–512.

Heath, K. M. and C. Hadley (1998) 'Dichotomous Male Reproductive Strategies in a Polygynous Human Society: Mating versus Parental Effort', *Current Anthropology* 39 (3): 369–74.

Hobbes, Thomas (1642/1983) *The Clarendon Edition of the Works of Thomas Hobbes, Volume 2: De Cive*, ed. Howard Warrender. Cambridge: Clarendon Press.

(1651) *Leviathan: Or the Matter and Power of a Commonwealth Ecclesiastical and Civil*. London: Andrew Crooke.

Hodge, Jonathan (2011) 'Darwinism after Mendelism: The Case of Sewall Wright's Intellectual Synthesis in His Shifting Balance Theory of Evolution (1931)', *Studies in History and Philosophy of Biological and Biomedical Sciences* 42 (1): 30–9.

Hoppensteadt, F. C. (1982) *Mathematical Methods of Population Biology*. Cambridge: Cambridge University Press.

Hrdy, Sarah Blaffer (1979) 'Infanticide among Animals: A Review, Classification, and Examination of the Implications for the Reproductive Strategies of Females', *Ethology and Sociobiology* 1 (1): 13–40.

(1981) *The Women That Never Evolved*. Cambridge, MA: Harvard University Press.

(1997) 'Raising Darwin's Consciousness: Female Sexuality and the Prehominid Origins of Patriarchy', *Human Nature* 8 (1): 1–49.

(2000) *Mother Nature: Maternal Instincts and How They Shape the Human Species*. New York: Ballantine Books.

(2011) *Mothers and Others: The Evolutionary Origins of Mutual Understanding*. Cambridge, MA: Belknap Press.

Hume, David (1738/1960) *Treatise of Human Nature*, edited by L. A. Selby-Bigge. Oxford: Clarendon Press.

(1748/1963) *An Enquiry Concerning Human Understanding*, edited by Ernest C. Mossner. New York: Washington Square Press.

(1772/1994) 'Of the Original Contract', in *Political Essays*, edited by Knud Haakonssen, pp. 186–201. Cambridge: Cambridge University Press.

Hurst, Laurence D (1997) 'Evolutionary Theories of Genomic Imprinting', in Wolf Reik and Azim Surani (eds.), *Genomic Imprinting*, pp. 211–37. Oxford: Oxford University Press.

Huxley, T. H. (1894/1989) *Evolution and Ethics*. Princeton, NJ: Princeton University Press.

Jeffreys, Sheila (1997) *The Idea of Prostitution*. North Melbourne: Spinifex.

Jelenkovic, Aline, Yoon-Mi Hur, Reijo Sund et al. (2016). 'Genetic and Environmental Influences on Adult Human Height across Birth Cohorts from 1886 to 1994', *Elife 5*.

Jorden-Young, Rebecca M. (2010) *Brain Storm: The Flaws in the Science of Sex Differences*. Cambridge, MA: Harvard University Press.

Josephson, Anna Leigh, Jacob Ricker-Gilbert and Raymond J. G. M. Florax (2014) 'How Does Population Density Influence Agricultural Intensification and Productivity? Evidence from Ethiopia', *Food Policy* 48: 142–52.

Joyce, Richard (2007) *The Evolution of Morality*. Cambridge, MA: MIT Press.

Kasper, Claudia, Maddalena Vierbuchen, Ulrich Ernst et al. (2017) 'Genetics and Developmental Biology of Cooperation', *Molecular Ecology* 26 (17): 4364–77.

Kavanagh, Patrick H., Bruno Vilela, Hannah J. Haynie et al. (2018) 'Hindcasting Global Population Densities Reveals Forces Enabling the Origin of Agriculture', *Nature Human Behaviour* 2 (7): 478–84.

Kingstone, Alan, Daniel Smilik and John D. Eastwood (2008) 'Cognitive Ethology: A New Approach for Studying Human Cognition', *British Journal of Psychology* 99 (3): 317–40.

Kitajima, Y. and Y. Ono (2016) 'Estrogens Maintain Skeletal Muscle and Satellite Cell Functions', *Journal of Endocrinology* 229: 267–75.

Kitcher, Philip (1985) *Vaulting Ambition: Soiobiology and the Quest for Human Nature*. Cambridge, MA: MIT Press.

Killian, J. Keith, James C. Byrd, James V. Jirle et al. (2000) 'M6P/IGF2 R Imprinting Evolution in Mammals', *Molecular Cell* 5: 707–16.

Kokko, Hanna, Andrés Lopez-Sepulcre, Lesley J. Morrell, Donald L. DeAngelis and Nicolas Perrin (2006) 'From Hawks and Doves to Self-Consistent Games of Territorial Behavior', *The American Naturalist* 167 (6): 901–12.

Kummer, Hans (1978) 'On the Value of Social Relationships to Nonhuman Primates: A Heuristic Scheme', *Social Science Information* 17 (4–5): 687–705.

Langergraber, Kevin E., John C. Mitani, David P. Watts and Linda Vigilant (2013) 'Male–Female Socio-spatial Relationships and Reproduction in Wild Chimpanzees', *Behavioral Ecology and Sociobiology* 67 (6): 861–73.

Law-Smith, M. J., D. I. Perrett, B. C. Jones, R. E. Cornwell, F. R. Moore and D. R. Feinberg (2006) 'Facial Appearance Is a Clue to Oestrogen Levels in

Women', *Proceedings of the Royal Society of London B: Biological Sciences* 273 (1583): 135–40.

Leiu, Judith M. (2013) 'What Did Women Do for the Early Church? The Recent History of a Question', *Studies in Church History* 49: 261–81.

Lepers, Romuald (2019) 'Sex Difference in Triathlon Performance', *Frontier in Physiology* 10: 973.

Lessnoff, Michael (1986) *Social Contract*. Atlantic Highlands, NJ: Humanities Press International.

Levitzky, Susan and Robyn Cooper (2000) 'Infant Colic Syndrome: Maternal Fantasies of Aggression and Infanticide', *Clinical Pediatrics* 39 (7): 395–400.

Leys, Sally P. (2015) 'Elements of a "Nervous System" in Sponges', *The Journal of Experimental Biology* 218 (Pt. 4): 581–91.

Locke, John (1690/1980) *Second Treatise of Government*, edited by C. B. Macpherson. Indianapolis, IN: Hackett.

Lukas, D. and T. H. Clutton-Brock (2013) 'The Evolution of Social Monogamy in Mammals', *Science* 341 (6145): 526–30.

Lumsden, Charles J. and Edward O. Wilson (1981) *Genes, Mind and Culture*. Cambridge, MA: Harvard University Press.

Lynch, Michael and Bruce Walsh (1998) *Genetics and Analysis of Quantitative Traits*. Sunderland, MA: Sinauer Associates.

Martin, J. S., E. J. Ringen, P. Duda and A. V. Jaeggi (2020) 'Harsh Environments Promote Alloparental Care Across Human Societies', *Proceedings of the Royal Society B: Biological Sciences* 287 (1933): 1–9.

Maynard Smith, John (1964) 'Group Selection and Kin Selection', *Nature* 201 (4924): 1145–7.

(1965) 'Prof. J. B. S. Haldane, F.R.S.', *Nature* 206 (4981): 239–40.

(1974a) 'The Theory of Games and the Evolution of Animal Conflicts', *Journal of Theoretical Biology* 47 (1): 209–21.

(1974b) *Models in Ecology*. Cambridge: Cambridge University Press.

(1975) 'Survival through Suicide', *New Scientist* 67 (964): 496–7.

(1982) *Evolution and the Theory of Games*. Cambridge: Cambridge University Press.

Maynard Smith, John and George R. Price (1973) 'The Logic of Animal Conflict', *Nature* 246 (5427): 15–18.

Maynard Smith, John and Eörs Szathmáry (1995) *Major Transitions in Evolution,* New York: W. H. Freeman Spektrum.

McElreath, Richard and Robert Boyd (2007) *Mathematical Models of Social Evolution*. Chicago, IL: University of Chicago Press.

McEvoy, Chad Joseph (2002) 'A Consideration of Human Xenophobia and Ethnocentrism from a Sociobiological Perspective', *Human Rights Review* 3 (3): 39–49.

McGee, Harold (2004) *On Food and Cooking: The Science and Lore of the Kitchen*. New York: Scribner.

MacKinnon, Catherine (1987) *Feminism Unmodified*. Cambridge, MA: Harvard University Press.

(1989) *Toward a Feminist Theory of the State*. Cambridge, MA: Harvard University Press.

McNamara, John M. (2013) 'Game Theory and Behavior', in Jonathan B. Losos, David A. Baum, Douglas J. Futuyma et al. (eds.), *The Princeton Guide to Evolution*, pp. 626–33. Princeton, NJ: Princeton University Press.

Miller, Goeffrey F., Joshua Tyber and Brent Jordan (2007) 'Ovulatory Cycle Effects on Tip Earnings by Lap-Dancers: Economic Evidence for Human Estrus?', *Evolution and Human Behavior* 28 (6): 375–81.

Miller, Saul L. and Jon K. Maner (2010) 'Scent of a Woman: Men's Testosterone Responses to Olfactory Ovulation Clues', *Psychological Science* 21 (2): 267–83.

Mochizuki, Atsushi, Takeda Yasuhiko and Iwasa Yoh (1996) 'The Evolution of Genomic Imprinting', *Genetics* 144 (3): 1283–95.

Moore, G. E. (1903) *Principia Ethica*. Cambridge: Cambridge University Press.

Mora-Garcia, Santiago and Justin Goodrich (2000) 'Genomic Imprinting: Seeds of Conflict', *Current Biology* 10 (2): R71–4.

Muller, Martin N. and Richard W. Wrangham (eds.) (2009) *Sexual Coercion in Primates and Humans: An Evolutionary Perspective on Male Aggression Against Females*. Cambridge, MA: Harvard University Press.

Murphy, Mark, (2019) 'The Natural Law Tradition in Ethics', in Edward N. Zalta (ed.), *The Stanford Encyclopedia of Philosophy*. https://plato.stanford.edu/archives/sum2019/entries/natural-law-ethics/.

Narveson, Jan (1988) *The Libertarian Idea*. Philadelphia: Temple University Press.

Nitecki, Doris V. and Matthew H. Nitecki (eds.) (1993) *Evolutionary Ethics*. Albany: State University of New York Press.

Nowak, Martin and Karl Sigmund (1993) 'A Strategy of Win-Stay, Lose-Shift That Outperforms Tit-for-Tat in Prisoner's Dilemma Game', *Nature* 364 (6432): 56–8.

(1994) 'The Alternating Prisoner's Dilemma', *Journal of Theoretical Biology* 168 (2): 219–26.

Nowak, Martin A., Corina E. Tarnita and Tibor Antal (2010) 'Evolutionary Dynamics in Structured Populations', *Philosophical Transactions of the Royal Society of London B: Biological Sciences* 365 (1537): 19–30.

Öhman, Arne and Susan Mineka (2001) 'Fears, Phobias, and Preparedness: Toward an Evolved Module of Fear and Fear Learning', *Psychological Review* 108 (3): 483–522.

O'Neill, Michael J., Robert S. Ingram, Paul B. Vrana and Shirley M. Tilghman (2000) 'Allelic Expression of Igf2 in Marsupials and Birds', *Development Genes and Evolution* 210 (1): 18–20.

Otto, Sarah P. and Troy Day (2007) *A Biologist's Guide to Mathematical Modelling in Ecology and Evolution.* Princeton, NJ: Princeton University Press.

Otto, Sarah P., Maria R. Servedio and Scott L. Nuismer (2008) 'Frequency-Dependent Selection and the Evolution of Assortative Mating', *Genetics* 179 (4): 2091–112.

Pagels, Elaine (1989) *Adam, Eve, and the Serpent: Sex and Politics in Early Christianity.* New York: Vintage Press.

Parfit, D. (2011) *On What Matters.* Oxford: Oxford University Press.

Peck, Steven L., Stephen P. Ellner and Fred Gould (1998) A Spatially Explicit Stochastic Model Demonstrates the Feasibility of Wright's Shifting Balance Theory', *Evolution* 52 (6): 1834–9.

Perkins, Jessica M., S. V. Subramanian, George Davey Smith and Emre Özaltin (2016) 'Adult Height, Nutrition and Population Health', *Nutrition Reviews* 74 (3): 149–65.

Perry, Gretchen and Martin Daly (2017) 'A Model Explaining the Matrilateral Bias in Alloparental Investment', *Proceeding of the National Academy of Science* 114 (35): 9290–5.

Pettitt, Paul (2018) 'The Rise of Modern Humans', in Chris Scarre (ed.), *The Human Past: World Prehistory and the Development of Human Societies* (4th ed.), pp. 108–48. London: Thames & Hudson.

Pigeon, Charles (2019) 'No-Ought-From-Is, the Naturalistic Fallacy, and the Fact/Value Distinction: The History of a Mistake', in Neil Sinclair (ed.), *The Naturalistic Fallacy*, pp. 71–95. Cambridge: Cambridge University Press.

Price, George (1969) 'Antlers, Intraspecific Combat, and Altruism' (Unpublished paper, British Library, George Price Correspondence, 1–32).

Rabin, Matthew (1993) 'Incorporating Fairness into Game Theory and Economics', *American Economic Review* 83 (5): 1281–302.

Ramsey, Grant (2013) 'Human Nature in a Post-essentialist World', *Philosophy of Science* 80 (5): 983–93.

Rawls, John (1958) 'Justice As Fairness', *The Philosophical Review* 67 (2): 164–94.

(1971) *A Theory of Justice*. Cambridge, MA: Harvard University Press.

Renshaw, Eric (1993) *Modelling Biological Populations in Space and Time*. Cambridge: Cambridge University Press.

Richards, Richard A. (2005) 'Evolutionary Naturalism and the Logical Structure of Valuation: The Other Side of Error Theory', *Cosmos and History: The Journal of Natural and Social Philosophy* 1 (2): 270–94.

Richerson, Peter J., Robert T. Boyd and Joseph Henrich (2003) 'Cultural Evolution of Human Cooperation', in Peter Hammerstein (ed.), *Genetic and Cultural Evolution of Cooperation*, pp. 357–88. Cambridge, MA: MIT Press.

Riedman, Marianne L. (1982) 'The Evolution of Alloparental Care and Adoption in Mammals and Birds', *The Quarterly Review of Biology* 57 (4): 405–35.

Ripstein, Arthur (2004) 'Authority and Coercion', *Philosophy and Public Affairs* 31 (1): 1–34.

(2006) 'Beyond the Harm Principle', *Philosophy and Public Affairs* 34 (3): 215–45.

Rood, Jon P. (1990) 'Group Size, Survival, Reproduction, and Routes to Breeding in Dwarf Mongooses', *Animal Behavior* 39 (3): 566–72.

Rowe, Locke, Erin Cameron and Troy Day (2005) 'Escalation, Retreat, and Female Indifference As Alternative Outcomes of Sexually Antagonistic Coevolution', in David Hosken and Rhonda Snook (eds.), *How Important Is Sexual Conflict?* Special Issue, *The American Naturalist* 165 (S5): S5–S18.

Ruse, Michael and Edward O. Wilson (1986) 'Moral Philosophy As Applied Science', *Philosophy* 61 (236): 173–92.

Russell, Andrew M. (2004) 'Mammals: Comparisons and Contrasts', in Walter D. Koenig and Janis L. Dickinson (eds.), *Ecology and Evolution of Cooperative Breeding in Birds*, pp. 210–27. Cambridge: Cambridge University Press.

Ryan, Joseph F. and Marta Chiodin (2015) 'Where Is My Mind? How Sponges and Placozoans May Have Lost Neural Cell Types', *Philosophical Transactions of the Royal Society B: Biological Sciences* 370 (1684): 1–6.

Ryan, Joseph F., Kevin Pang, Christine E. Schnitzler et al. (2013) 'The Genome of the Ctenophore *Mnemiopsis leidyi* and Its Implications for Cell Type Evolution', *Science* 342 (6164): 1336–44.

Sahlins, Marshall. (1976) *The Uses and Abuses of Biology*. Ann Arbor: University of Michigan Press.

Sayre-McCord, Geoffrey (2013) 'Hume and Smith on Sympathy, Approbation, and Moral Judgment', *Social Philosophy and Policy* 30 (1-2): 208–36.

Scanlon, Thomas M. (1998) *What We Owe Each Other*. Cambridge, MA: Harvard University Press.

Scarre, Chris (ed.) (2018) *The Human Past: World Prehistory and the Development of Human Societies* (4th ed.). London: Thames & Hudson.

Schlegel, Rolf H. J. (2018) *History of Plant Breeding*. New York: Routledge.

Scott, George Ryley (2013) *A History of Prostitution from Antiquity to the Present Day*. London: Routledge.

Silk, Joan B. (1980) 'Adoption and Kinship in Oceania', *American Anthropologist* 82 (4): 799–820.

(2007) 'The Adaptive Value of Sociality in Mammalian Groups', *Philosophical Transactions of the Royal Society B: Biological Sciences* 362 (1480): 539–59.

Simpson, George Gaylord (1953) 'The Baldwin Effect', *Evolution* 7 (2): 110–17.

Sinclair, Neil (ed.) (2019) *The Naturalistic Fallacy*. Cambridge: Cambridge University Press.

Skidelsky, Robert (2020) *What's Wrong with Economics? A Primer for the Perplexed*. New Haven, CT: Yale University Press.

Skyrms, Brian (1996) *Evolution of the Social Contract*. Cambridge: Cambridge University Press.

Smuts, Barbara (1992) 'Male Aggression Against Women: An evolutionary Perspective', *Human Nature* 3 (1): 1–44.

(1995) 'The Evolutionary Origins of Patriarchy', *Human Nature* 6 (1): 1–32.

Smuts, Barbara and Robert Smuts (1993) 'Male Aggression and Sexual Coercion of Females in Nonhuman Primates and Other Mammals: Evidence and Theoretical Implications', in Peter Slater, Manfred Milinski, Charles Snowden and Jay Rosenblatt (eds.), *Advances in the Study of Behavior*, Vol. 22, pp. 1–63. Cambridge, MA: Academic Press.

Sober, Elliott (1994) *From a Biological Point of View*. Cambridge: Cambridge University Press.

Sober, Elliott and David Sloan Wilson (1998) *Unto Others: The Evolution and Psychology of Unselfish Behaviour*. Cambridge, MA: Harvard University Press.

Sousa, André M. M., Kyle A. Meyer, Gabriel Santpere, Forrest O. Gulden and Nenad Sestan (2017) 'Evolution of the Human Nervous System Function, Structure, and Development', *Cell* 170 (2): 226–47.

Spencer, Hamish G., Marcus W. Feldman and Andrew G. Clark (1998) 'Genetic Conflicts, Multiple Paternity and the Evolution of Genomic Imprinting', *Genetics* 148 (2): 893–904.

Spencer, Herbert (1851) *Social Statics: Or, the Conditions Essential to Human Happiness Specified, and the First of Them Developed*. London: John Chapman.

Stephens, Christopher (1996) 'Modelling Reciprocal Altruism', *British Journal for the Philosophy of Science* 47 (4): 533–51.

Stephensen, Charles B. (1999) 'Burden of Infection on Growth Failure', *The Journal of Nutrition* 129 (2): 534–8.

Stumpf, R. M. and C. Boesch (2010) 'Male Aggression and Sexual Coercion in Wild West African Chimpanzees, *Pan troglodytes verus*', *Animal Behaviour* 79 (2): 333–42.

Sturgeon, Nicholas L. (2003) 'Moore on Ethical Naturalism', *Ethics* 113 (3): 528–56.

Sumner, L. Wayne (1989) *The Moral Foundation of Rights*. Oxford: Clarendon Press.

(2004) *The Hateful and the Obscene: Studies in the Limits of Free Expression*. Toronto: University of Toronto Press.

Swedell, Larissa, Liane Leedom, Julian Saunders and Mathew Pines (2014) 'Sexual Conflict in a Polygynous Primate: Costs and Benefits of a Male-Imposed Mating System', *Behavioral Ecology and Sociobiology* 68 (2): 263–73.

Tarnita, Corina E., Hisashi Ohtsuki, Tibor Antal, Feng Fu and Martin A. Nowak (2009) 'Strategy Selection in Structured Populations', *Journal of Theoretical Biology* 259 (3): 570–81.

Thompson, R. Paul (1990) 'Evolutionary Ethics, Darwinian Ethics and Ethical Naturalism', *Human Evolution* 5: 133–8.

(ed.) (1995) *Issues in Evolutionary Ethics*. Albany: State University of New York Press.

(1999) 'Evolutionary Ethics: Its Origins and Contemporary Face', *Zygon* 34 (3): 473–585.

(2002) 'The Evolutionary Biology of Evil', *The Monist* 85 (2): 239–59.

(2007) 'Formalisations of Evolutionary Biology', in Mohan Matthen and Christopher Stephens (eds.), *Handbook of the Philosophy of Science, Volume 2: Philosophy of Biology*, pp. 485–523. New York: Elsevier.

(2011) 'Theories and Models in Medicine', in Frederick Gifford (ed.), *Handbook of the Philosophy of Science, Volume 4: Philosophy of Medicine*, pp. 115–36. New York: Elsevier.

(2014) 'Darwin's Theory and the Value of Mathematical Formalisation', in R. Paul Thompson and Denis Walsh (eds.), *Evolutionary Biology: Conceptual, Ethical and Religious Issues*, pp. 109–36. Cambridge: Cambridge University Press.

Toth, Nicholas and Kathy Schick (2018) 'African Origins', in Chris Scarre (ed.), *The Human Past: World Prehistory and the Development of Human Societies* (4th ed.), pp. 46–70. London: Thames & Hudson.

Trigger, Bruce G. (2003) *Understanding Early Civilizations*, Cambridge: Cambridge University Press.

Trivers, Robert, L. (1971) 'The Evolution of Reciprocal Altruism', *Quarterly Review of Biology*, 46 (1): 35–57.

(1972) 'Parental Investment and Sexual Selection', in B. Campbell (ed.), *Sexual Selection and the Descent of Man*, pp. 136–79. London: Heinemann.

(1974) 'Parent-Offspring Conflict', *American Zoologist* 14 (1): 249–64.

Vannelli, Ron (2015) *The Evolution of Human Sociability: Desires, Fears, Sex and Society*. Cambridge: Cambridge University Press.

Van Valen, Leigh (1973) 'A New Evolutionary Law', *Evolutionary Theory* 1: 1–30.

van Veelen, Matthijs, Julián García, David G. Rand and Martin A. Nowak (2012) 'Direct Reciprocity in Structured Populations', *Proceedings of the National Academy of Sciences of the United States of America* 109 (25): 9929–34.

Vigne, Jean-Denis (2011) 'The Origins of Animal Domestication and Husbandry: A Major Change in the History of Humanity and the Biosphere', *Comptes Rendus Biologies* 334 (3): 171–81.

Vitzthum, Virginia J. (2008) 'Evolutionary Models of Women's Reproductive Functioning', *Annual Review of Anthropology* 37 (1): 53–73.

von Neumann, John and O. Morgenstern (1944) *Theory of Games and Economic Behavior*. Princeton, NJ: Princeton University Press.

Voorhees, Burton, Dwight Read and Liane Gabora (2020) 'Identity, Kinship, and the Evolution of Cooperation', *Current Anthropology* 61 (2): 194–218.

Waal, Frans B. M. de (1997) *Good Natured: The Origins of Right and Wrong in Humans and Other Animals*. Cambridge, MA: Harvard University Press.

Waddington, Conrad H. (1957) *The Strategy of the Genes*. London: Allen & Unwin.

Wade, Michael J., and Charles J. Goodnight (1998) 'The Theories of Fisher and Wright in the Context of Metapopulations: When Nature Does Many Small Experiments', *Evolution* 52 (6): 1537–53.

Weitzer, Ronald (2012) *Legalizing Prostitution: From Illicit Vice to Lawful Business*. New York: New York University Press.

Wetzel, James (2012) 'Augustine on the Will', in Mark Vessey (ed.), *A Companion to Augustine*, pp. 337–52. Oxford: Wiley-Blackwell.

Williams, George C. (1966) *Adaptation and Natural Selection: A Critique of Some Evolutionary Thought*. Princeton, NJ: Princeton University Press.

Wilson, Edward O. (1978) *On Human Nature*. Cambridge, MA: Harvard University Press.

(1975) *Sociobiology: The New Synthesis*. Cambridge, MA: Harvard University Press.

Wood, Anew R., Tonu Esko, Jian Yang et al. (2014) 'Defining the Role of Common Variation in the Genomic and Biological Architecture of Adult Human Height', *Nature Genetics* 46 (11): 1173.

Wood, W. Jay (2000) 'What Would Augustine Say?', *Christian History* 19 (3): 36–8.

Wright, Sewall (1922) 'Coefficients of Inbreeding and Relationship', *The American Naturalist* 56 (645): 330–8.

(1931) 'Evolution in Mendelian Populations', *Genetics* 16 (2): 97–159.

(1932) 'The Roles of Mutation, Inbreeding, Crossbreeding and Selection in Evolution', *Proceedings of the Sixth Annual Congress of Genetics* 1: 356–66.

(1969) *Evolution and the Genetics of Populations, Volume 2: The Theory of Gene Frequencies*. Chicago, IL: University of Chicago Press.

(1977) *Evolution and the Genetics of Populations, Volume 3: Experimental Results and Evolutionary Deductions*. Chicago, IL: University of Chicago Press.

Zeh, David W. and Jeanne A. Zeh (2000) 'Reproductive Mode and Speciation: The Viviparity-driven Conflict Hypothesis', *BioEssays* 22 (10): 938–46.

(2002) 'Maternal-Fetal Conflict', in Mark Pagel (ed.), *Encyclopedia of Evolution*. Oxford: Oxford University Press. https://doi.org/10.1093/acref/9780195122008.001.0001

Zhou, W.-X., D. Sornette, R. A. Hill and R. I. M. Dunbar (2005) 'Discrete Hierarchical Organization of Social Group Sizes', *Proceedings of the Royal Society of London*, 272B (1561): 439–44.

For Jennifer Christine McShane
Dum vivimus, vivamus
Nuair a bhíonn súile na hÉireann ag miongháire

Cambridge Elements ⹀

Philosophy of Biology

Grant Ramsey

KU Leuven

Grant Ramsey is a BOFZAP research professor at the Institute of Philosophy, KU Leuven, Belgium. His work centers on philosophical problems at the foundation of evolutionary biology. He has been awarded the Popper Prize twice for his work in this area. He also publishes in the philosophy of animal behavior, human nature and the moral emotions. He runs the Ramsey Lab (theramseylab.org), a highly collaborative research group focused on issues in the philosophy of the life sciences.

Michael Ruse

Florida State University

Michael Ruse is the Lucyle T. Werkmeister Professor of Philosophy and the Director of the Program in the History and Philosophy of Science at Florida State University. He is Professor Emeritus at the University of Guelph, in Ontario, Canada. He is a former Guggenheim fellow and Gifford lecturer. He is the author or editor of over sixty books, most recently *Darwinism as Religion: What Literature Tells Us about Evolution*; *On Purpose*; *The Problem of War: Darwinism, Christianity, and their Battle to Understand Human Conflict*; and *A Meaning to Life*.

About the Series

This Cambridge Elements series provides concise and structured introductions to all of the central topics in the philosophy of biology. Contributors to the series are cutting-edge researchers who offer balanced, comprehensive coverage of multiple perspectives, while also developing new ideas and arguments from a unique viewpoint.

Cambridge Elements $\equiv$

Philosophy of Biology

Elements in the Series

A full series listing is available at www.cambridge.org/EPBY

Printed in the United States
by Baker & Taylor Publisher Services